AF595558

Achieving the Circular Economy

Achieving the Circular Economy: Exploring the Role of Local Governments, Business and Civic Society in an Urban Context

Editors

Jenny Palm
Nancy Bocken

MDPI • Basel • Beijing • Wuhan • Barcelona • Belgrade • Manchester • Tokyo • Cluj • Tianjin

Editors
Jenny Palm
Lund University
Sweden

Nancy Bocken
Lund University
Sweden

Editorial Office
MDPI
St. Alban-Anlage 66
4052 Basel, Switzerland

This is a reprint of articles from the Special Issue published online in the open access journal *Energies* (ISSN 1996-1073) (available at: https://www.mdpi.com/journal/energies/special_issues/achieving_circular_economy).

For citation purposes, cite each article independently as indicated on the article page online and as indicated below:

LastName, A.A.; LastName, B.B.; LastName, C.C. Article Title. *Journal Name* **Year**, *Volume Number*, Page Range.

ISBN 978-3-0365-0118-5 (Hbk)
ISBN 978-3-0365-0119-2 (PDF)

Contents

About the Editors

Jenny Palm is a professor in sustainable urban governance for a transition to low-carbon and resource-efficient economies at Lund University. Jenny Palm has a background in energy systems research where she conducts interdisciplinary research combining sociotechnical systems theory with an analysis of planning process, governance, technology diffusion and end-users. The objects of study are urban infrastructure and planning, sustainable city districts, prosumers, grassroot initiatives and community energy. Palm's projects are conducted in collaboration with one or several actors outside academia. Palm is the theme leader for the research group Urban Governance and Experimentation at IIIEE which encompasses around 10 researchers.

Nancy Bocken is a professor in sustainable business at Maastricht University. Bocken's main areas of interest around sustainability are business models, business experimentation, innovation for sustainability, scaling up sustainable businesses and closing the 'idea–action' gap in sustainability. She has led several international collaborative research projects on sustainable business models. Bocken has been a professor at Lund University and has supervised and taught at the University of Cambridge Institute for Sustainability Leadership, the Cambridge Department of Engineering and Judge Business School. In 2013, she was a visiting fellow at Yale University, where she was lecturing on the Corporate Environmental Strategy course, while continuing her work on sustainable business models. Interwoven with her academic work, Nancy regularly advises a range of organisations across sectors and continents, and various sustainability start-up initiatives. She co-founded HOMIE, who develop and test circular business models, starting with pay per use home appliances.

Preface to "Achieving the Circular Economy: Exploring the Role of Local Governments, Business and Civic Society in an Urban Context"

Urbanisation and climate change are pushing cities to find novel pathways leading to a sustainable future. The urban context may be viewed as a new experimentation space to accelerate the transition to a circular economy. The contributions of this Special Issue give a fantastic overview of the experimental capacity of the city and we wish to express our gratitude to the authors for sharing their findings and reflections in this Special Issue. The Special Issue includes interdisciplinary papers on a variety of topics and sectors: governing modes, textile industry, repair ecosystems for mobile phones, waste management, food producers and farmer markets, citizens and climate change. The geographical focus was mainly Europe, but China and the US were also studied. The authors of this Special Issue have illuminated the multitude of topics, methods, tools and perspectives that need to be included when exploring how circular cities can evolve today and in the future. We want to thank all the authors for their contributions.

Jenny Palm, Nancy Bocken
Editors

Editorial

Achieving the Circular Economy: Exploring the Role of Local Governments, Business and Citizens in an Urban Context

Jenny Palm [1,*] and Nancy Bocken [1,2]

[1] International Institute for Industrial Environmental Economics (IIIEE), Lund University, P.O. Box 196, SE-221 00 Lund, Sweden; nancy.bocken@maastrichtuniversity.nl

[2] Maastricht Sustainability Institute, School of Business and Economics, Maastricht University, Tapijn 11 Building D, P.O. Box 616, 6200 MD Maastricht, The Netherlands

* Correspondence: jenny.palm@iiiee.lu.se; Tel.: +46-46-222-02-42

Received: 28 January 2021; Accepted: 4 February 2021; Published: 8 February 2021

1. Objectives of This Special Issue

The urban context is an experimentation space to accelerate the transition to a circular economy. This Special Issue explores how and why cities engage in circularity. This Special Issue includes papers from different regions, on a variety of topics, using different methods and suggesting a multitude of tools. In sum, together they show the complexity that cities are facing, but they also provide insights into how circular cities may evolve in the future.

2. Overview of the Papers Included

Table 1 gives an overview of included articles and which scope, geographical focus, methods and tools the articles discuss.

The included papers cover a wide variety of topics and actors: governing modes, the textile industry, repair ecosystems for mobile phones, waste management, food producers and farmers markets, and citizens and climate change. The geographical focus is EU, USA and China. The research methods applied in the papers are also diversified, including both quantitative and qualitative approaches. Even if all papers discuss different challenges that cities and the circular economy are facing, they all end up being solution-oriented, with suggestions for tools to apply in the future. Below, the different papers are presented in more depth.

In the paper of Palm, Södergren and Bocken, "The Role of Cities in the Sharing Economy: Exploring Modes of Governance in Urban Sharing Practices" [1], the potential roles cities might have in governing the sharing economy is explored. Cities have provided space for various sharing initiatives such as car sharing, public libraries and repair workshops. The potential governing roles of cities in the sharing economy are discussed in relation to the experience in four Swedish cities: Stockholm, Gothenburg, Malmö and Umeå. Three dominant modes of governing were identified, namely: governing by provision and authority; governing by partnership and enabling; and governing through volunteering. The four studied cities applied all three governing modes, although with a primary focus on governing by authority and governing through partnership. When it comes to sharing projects characterized by governing through volunteering, it was always the local government initiating these projects, even if these projects were formally run by a non-governmental organisation '(NGO). One important conclusion from this paper is that it is important that cities reflect upon their powerful position, and, that in the case the city becomes too dominant, it might out-compete both businesses and initiatives from NGOs.

Table 1. Papers in this Special Issue.

Authors	Title	Scope/Actors	Country Focus	Method(s)	Tool(s)/Outcomes
Palm, Södergren and Bocken	The Role of Cities in the Sharing Economy: Exploring Modes of Governance in Urban Sharing Practices	Governing modes, local governments, cities	Sweden	Workshops, semi-structured interviews and participant observation	Sharing solutions
Wang, Chen, Cheng, Zhou, Li and Yang	Factorial Decomposition of the Energy Footprint of the Shaoxing Textile Industry	Textile industry	China	Energy footprint (EFP); Logarithmic Mean Divisia Index (LMDI) Decomposition Model	Energy footprint (EFP) indicator
Türkeli, Huang, Stasik and Kemp	Circular Economy as a Glocal Business Activity: Mobile Phone Repair in the Netherlands, Poland and China	Repair ecosystems for mobile phones in countries	Netherlands, Poland and China	Questionnaire, interviews, comparison	Insight into the repair sector in The Netherlands, China, and Poland and policy advice
Palafox-Alcantar, Hunt and Rogers	A Hybrid Methodology to Study Stakeholder Cooperation in Circular Economy Waste Management of Cities	Different stakeholders in waste management	UK	Scenario Analysis; Multi-Criteria Decision Analysis; and Game Theory	Hybrid methodology encouraging cooperation in decision-making processes
Nogueira, Ashton, Teixeira, Lyon and Pereira	Infrastructuring the Circular Economy	Food producers and farmers markets	USA	Participatory action research, co-creation, material and energy flow analysis, life cycle assessment, network analysis, systems dynamics modeling, design methods.	Innovation model and four I's model
Davidescu, Apostu and Paul	Exploring Citizens' Actions in Mitigating Climate Change and Moving toward Urban Circular Economy. A Multilevel Approach	Citizens and climate change	All EU member states	Multi-Level Econometric Modelling; logistic regression analysis	Encouraging citizens to take action
Zaleski and Chawla	Circular Economy in Poland: Profitability Analysis for Two Methods of Waste Processing in Small Municipalities	Waste processing, incineration and torrefaction	Poland	Case study, scenarios, comparison, profitability analysis	Torrefaction (Torrefaction changes biomass properties to provide a better fuel quality for combustion and gasification)

Wang, Chen, Cheng, Zhou, Li and Yang's paper, "Factorial Decomposition of the Energy Footprint of the Shaoxing Textile Industry" [2], calculates the energy footprint (EFP) from textile production using Shaoxing's textile industry as a case study. The analysis focuses on the relationship between economic growth and environmental pressure. The analysis showed that from 2005 to 2018, the EFP first increased and then decreased. The authors conclude that Shaoxing's textile companies could accelerate the production of ecological textiles and promote clean production, printing and dyeing technologies. The local government could, together with social investors, contribute by initiating an innovation fund. Another suggestion by the authors is that Shaoxing's government should use their authority and close down printing and dying companies that consume significant amounts of energy and generate high levels of emissions. The authors also indicate the importance of developing policies with the aim to promote high value-added textile product manufacturing with less energy intensity.

Türkeli, Huang, Stasik and Kemp's article, titled "Circular Economy as a Glocal Business Activity: Mobile Phone Repair in the Netherlands, Poland and China" [3], looks into the repair of mobile phones. Repairing mobile phones is an excellent example of circular economy in an urban setting. This repair business does not only extend the lifetime of the phone, but also reduces the need for virgin materials used for constructing new phones. The authors embed the study in earlier research on firm level competitiveness and closed-loop design through repair. The focus for the analysis is the business ecosystem of independent mobile phone repair shops in three countries, the Netherlands, Poland and China, where questionnaires have been sent out to repair shops. The findings show that maintaining business' direct contact with customers is vital to sustain trust. The challenges the repair shops faced differed between the countries. In China, high cost for spare parts and low prices on new mobile phones were central, while the Netherlands and Poland experienced big challenges with competition from informal repair activities and new repair shops.

Palafox-Alcantar, Hunt and Rogers, in their paper titled "A Hybrid Methodology to Study Stakeholder Cooperation in Circular Economy Waste Management of Cities" [4], study the waste management of cities from a circular economy perspective and discuss how efficient waste management process resources can be fed back into the consumption process rather than reach an end-of-life. The contribution of the paper is to identify how collaboration can be engendered using a hybrid Game Theory approach, including scenario analysis and multi-criteria decision analysis. A case study of Birmingham in the UK is presented. The results show, for example, that cooperation needs to be embedded in circular economy adoption. A decision-making tool, such as the one presented in the paper, can be an important means to support this.

Nogueira, Ashton, Teixeira, Lyon and Pereira's paper, "Infrastructuring the Circular Economy" [5], discusses the need to reconfigure both hard (material and tangible aspects) and soft infrastructure (institutions, intangible aspects, and social behavior) to achieve efficient material resource cycling. The authors develop a new framework and a model including the range of resources organizations utilize when creating value for the organization, society, or the planet. Participatory action research methods are used in co-creation processes to synthesize knowledge on hard and soft infrastructures in relation to urban food producers and farmers markets in Chicago. The authors conclude that using a relational perspective gives insights into new opportunities for city interventions, where the different actors embedded in a situation become the means through which resources are mobilized and activated. By including dynamic interactions, cities can better understand how these interactions shape an infrastructural intervention for the circular economy and, by that, cities can also actively influence how to combine different types of resources to generate a sustainable transition.

Davidescu, Apostu and Paul focus on citizens in their paper titled "Exploring Citizens' Actions in Mitigating Climate Change and Moving toward Urban Circular Economy. A Multilevel Approach" [6]. The paper investigates why people engage in certain activities to contest climate change and choose to adopt more actions than others to mitigate climate change. The data used in the paper come from the cross-national dataset Eurobarometer and a study covering residents in all EU member states aged 15 years and over. In total, 27,655 individuals were interviewed, and of these, 18,529 were individuals from urban areas. The results show that climate change was perceived to be a very serious problem by most of the respondents. Many citizens declared that they have personally taken action to fight climate change; 25% of the respondents declared that they have taken more than five actions and 1% of the individuals declared that they have taken nine or more actions to mitigate climate change. The authors conclude that it is important to also consider citizens' attitudes towards climate change when developing strategies for circular cities.

Zaleski and Chawla, in their paper "Circular Economy in Poland: Profitability Analysis for Two Methods of Waste Processing in Small Municipalities" [7], discuss the implementation of the circular economy paradigm in Poland. Even if Poland has been successful in reducing the volume of generated waste during the last few decades, over 42% of waste is still being land-filled. In the article, profitability analysis is carried out for two methods of waste processing, namely incineration and torrefaction. The results show that torrefaction is a more desirable waste processing option as a step towards the implementation of the circular economy in the urban context. Torrefaction is also more profitable compared to incineration. Poland has so far not implemented any torrefaction plants on a large-scale basis for processing municipal waste, but this is something that Zalesk and Chawla recommend Poland to consider, which would be in line with the country's strategy to achieve a circular economy.

3. Conclusions

This Special Issue included interdisciplinary papers on a variety of topics and sectors: governing modes, the textile industry, repair ecosystems for mobile phones, waste management, food producers and farmers markets, and citizens and climate change. The geographical focus was mainly Europe, but China and the US were also studied. The authors of this Special Issue have illuminated a multitude of topics, methods, tools and perspectives that need to be included when exploring how circular cities can evolve today and in the future.

The lessons learned from the papers are manifold. First, cities have a multitude of tools to govern the circular economy. As shown by Wang et al., they can encourage circular initiatives by, for example, institutionalizing an innovation fund or using their authority to close down activities leading to major emissions. The potential positive role of city governments for the circular economy is undisputed, but Palm et al. showed that city governments also need to reflect upon their power and ensure that their strategies and goals do not outcompete civic society and their engagement. Second, the need for cooperation between actors was highlighted in several papers. A decision-making tool was suggested by Palafox-Alcantar et al., while Türkeli et al. emphasized the need for repair shops to establish direct contact with their customers and create trust. Nogueira et al. suggested that by embedding different actors (not only the usual suspects) in an activity, cities can actively contribute so that new cooperation and resource flows develop. Moreover, Davidescu et al. highlighted the need to also include citizens when developing strategies for circular cities. Third, the importance of reducing waste was also a recurring issue in the papers. Zaleski and Chawla focused on waste processing and found that torrefaction was more profitable than incineration and also a good choice for small cities striving to achieve a circular economy.

Finally, the papers covered a range of circular economy strategies, including repair to extend product lifetimes and slow resource loops; waste management approaches to recycle materials and close resource loops; and energy reduction per product and processes to narrow resource loops. By highlighting specific resource strategies, tools, methods and actors to involve, the papers contribute to our understanding of what a circular city could constitute and how it could emerge in the future. Yet, the papers are only starting to scratch the surface of the methods, approaches and collaborations needed to put the circular economy into practice. A diverse range of solutions and collaborations would need to co-exist to implement the circular economy in practice. To conclude, this Special Issue aims to inspire and give a solid base for further investigations of and experimentation with new solutions for circular cities.

Author Contributions: Both authors have contributed equally to this editorial. Both authors have read and agreed to the published version of the manuscript.

Funding: We would like to acknowledge the funding by the Swedish Energy Agency project no 46016-1 "Smart symbiosis—collaboration for common resource flows".

Institutional Review Board Statement: Not applicable.

Informed Consent Statement: Not applicable.

Data Availability Statement: Not applicable.

Conflicts of Interest: The authors declare no conflict of interest.

References

1. Palm, J.; Södergren, K.; Bocken, N. The Role of Cities in the Sharing Economy: Exploring Modes of Governance in Urban Sharing Practices. *Energies* **2019**, *12*, 4737. [CrossRef]
2. Wang, X.; Chen, X.; Cheng, Y.; Zhou, L.; Li, Y.; Yang, Y. Factorial Decomposition of the Energy Footprint of the Shaoxing Textile Industry. *Energies* **2020**, *13*, 1683. [CrossRef]
3. Türkeli, S.; Huang, B.; Stasik, A.; Kemp, R. Circular Economy as a Glocal Business Activity: Mobile Phone Repair in the Netherlands, Poland and China. *Energies* **2019**, *12*, 498. [CrossRef]
4. Palafox-Alcantar, P.G.; Hunt, D.V.; Rogers, C.D. A Hybrid Methodology to Study Stakeholder Cooperation in Circular Economy Waste Management of Cities. *Energies* **2020**, *13*, 1845. [CrossRef]
5. Nogueira, A.; Ashton, W.; Teixeira, C.; Lyon, E.; Pereira, J. Infrastructuring the Circular Economy. *Energies* **2020**, *13*, 1805. [CrossRef]
6. Davidescu, A.A.; Apostu, S.-A.; Paul, A. Exploring Citizens' Actions in Mitigating Climate Change and Moving toward Urban Circular Economy. A Multilevel Approach. *Energies* **2020**, *13*, 4752. [CrossRef]
7. Zaleski, P.; Chawla, Y. Circular Economy in Poland: Profitability Analysis for Two Methods of Waste Processing in Small Municipalities. *Energies* **2020**, *13*, 5166. [CrossRef]

Article

Circular Economy as a Glocal Business Activity: Mobile Phone Repair in the Netherlands, Poland and China

Serdar Türkeli [1,*], Beijia Huang [2], Agata Stasik [3] and René Kemp [1]

[1] School of Business and Economics, United Nations University-MERIT/Maastricht University, 6211AX Maastricht, The Netherlands; r.kemp@maastrichtuniversity.nl
[2] School of Environment and Architecture, University of Shanghai for Science and Technology, Shanghai 200093, China; beijia.huang@yale.edu
[3] Department of Management in Network Society, Koźmiński University, 03-301 Warsaw, Poland; astasik@kozminski.edu.pl
* Correspondence: turkeli@merit.unu.edu

Received: 24 December 2018; Accepted: 27 January 2019; Published: 5 February 2019

Abstract: Repair of mobile phones fits with the vision of a circular economy in an urban context and with the Sustainable Development Goal 11 Sustainable Cities and Communities. Drawing on the literature about firm level competitiveness and closed-loop design through repair, remanufacturing or recycling, we analyze the business ecosystem of independent mobile phone repair shops in the Netherlands, Poland and China as a glocal business activity. The analysis is based on primary data collection through a questionnaire to independent repair shops in the Netherlands (n = 130), Poland (n = 443) and China (n = 175) with response rates of 13%, 12%, 40%, respectively; and 17 interviews in the Netherlands, 40 in Poland, and 70 in China. Findings indicate that to maintain a strong position in the local market and to sustain the trust of customers, independent mobile phone repair shops offer a range of customized services based on direct contact with customers. In China, the increasing prices of spare parts and falling prices of mobile phones constitute the most important challenges, whereas in the Netherlands and Poland, the most important challenges are the competitive pressures from informal repair activities, and new repair shops. Our research also revealed that repairability strongly depends on the global manufacturers' circularity choices.

Keywords: circular economy; business ecosystem; glocality; mobile phone repair; the Netherlands; China; Poland

---

1. Introduction

Repair of mobile phones fits with the vision of a circular economy in an urban context. Repair as a local loop and an inner cycle of a circular economy is significantly distinguished from relatively larger and also potentially regional and global loops, the outer cycles of remanufacturing and recycling activities of a circular economy. Local repair businesses in an urban context extend the lifetime of mobile phones and reduce the need for virgin materials, which in theory results in environmental benefits [1–5]. In addition, local repair businesses contribute to lower levels of waste electrical and electronic equipment (WEEE) in cities and increase maximum possible usage of mobile phones over time, along with supporting refurbishment, remanufacturing, and recycling activities. Globally, repair activities and repair shops can also be viewed as local contributors to the global development agenda, namely, the Sustainable Development Goals (SDGs), which were agreed upon by the 193 members states of the United Nations (e.g., SDG 11 Sustainable Cities and Communities), and especially the target of "reducing the adverse per capita environmental impact of cities, including by paying special attention to air quality and municipal and other waste management by 2030" [6,7].

Repair activities constitute a business response to an economic opportunity, which complements civil society responses of product repair, and repair and resale activities of social enterprises (e.g., organized in Europe via RREUSE (RREUSE represents social enterprises active in re-use, repair and recycling in Europe)). Locally, mobile phone repair activities also offer employment to those working (migrant workers in the Netherlands, who dominate the business).

Concentrating on the historical and contemporary mobile phone technology, and its global-scale market, a hundred years ago Finnish inventor Eric Tigerstedt filed a patent for a pocket-size folding telephone with a very thin carbon microphone in 1917; in 2017 we observe that the number of mobile phone users is forecasted to reach 4.77 billion people around the world [8]. In 2017, as many as 1.54 billion mobile phones were sold worldwide [9]. In the Netherlands, mobile phone penetration rate reached 68.8% with 11.7 million (mln) users; in Poland, 66.5% with 25.6 mln users; while in China, its 51.7% penetration rate corresponds to 717.3 mln users [10]. New generation mobile phones, especially smartphones with internet-enabled, feature-rich applications, are relatively expensive devices with a mid-range retail price between EUR 150 and 450 [11]. Typically, these technological devices still come with relatively fragile screens (although improvements against fragility by manufactures are being introduced) or otherwise hard-to-replace components when/if they malfunction and/or require cosmetic corrections [12]. These financial and technological constraints encourage consumers more frequently to turn to repair shops to rescue their devices which are still operable, and to pursue less expensive alternatives compared to replacement and upgrading options. Initially, mobile phone users turned to official brand shops and to non-official shops for repair (people working from home) but, over time, independent repair shops were created, to which people could bring their phone for repair and special services, with the repairs being done quickly and at a far lower cost than when the repair is done by the suppliers working for official brands. Helped by low entry barriers, these independent repair shops developed strategies that offered competitive, and also superior, customer value.

Independent repair shops are growing in number worldwide, and the objective of this article is to investigate the business of mobile repair shops from a glocal business ecosystem perspective based on the diamond model of Porter in three different local contexts, namely, the Netherlands, Poland and China. The overarching research question of this article is: to what extent do mobile phone repair activities and circularity relations of independent repair shops differ in the Netherlands, Poland and China, and under which conditions do they try to competitively operate in their business ecosystems? With this article, we aim to fill this gap in the literature, by offering an analysis of 'the diamond' of firm strategy and offerings, factor conditions, demand conditions, and the links with upstream and downstream industries (especially parts producers) and circularity relations with recycling companies and remanufacturers.

The article proceeds as follows: in Section 2 we position our article in terms of related literature and provide its conceptual background: Porter's diamond model depicting the dynamic conditions affecting competitiveness of firms, the framework for a closed-loop design for repair, remanufacturing and recycling in the context of a circular economy, and the hierarchy of secondary market production processes with respect to labor content, performance, and warranty criteria. In Section 3, we provide details about the designed and implemented survey, interviews and the rationale for opting for using those methods. Section 4 presents the findings from Dutch, Chinese and Polish contexts. In Section 5, we further discuss our findings in the context of a circular economy and provide conclusions on future research and policy directions.

2. Literature Review and Conceptual Background

The behavior of consumers in turning to mobile phone repair shops or remanufactured mobile phones has been studied in detail by scholars [13–17], yet the emergence and heterogeneous evolution of mobile phone repair shops and the sector in transition as a whole (including demand side and circularity dynamics) from independent mobile phone repair shops' perspective is less studied, and accordingly less well known [17–20]. Comparative studies are even rarer.

The mobile phone repair industry and market are growing: Watson et al. report that "the repair industry is exploding in Nordic countries with phone repair shops appearing on the high streets of every market town" [19]. The situation is likewise observable in the Netherlands, Poland and China. Repair rates are higher in East Asia—66% in China and 64% in South Korea—than in Germany (23%) and in the US (28%) [21,22]. In the Netherlands, 12% of the currently used phones are either used (second-hand/directly sold) or refurbished, and the most common used/refurbished phones are Apple IOS phones (19%), followed by LG (12%), Sony and Samsung (each 9%) and Huawei (4%) [23]. In China, according to Kantar Worldpanel 2017 data for the mobile phone market, Android phones accounted for 71.2% of market share, while Apple IOS phones accounted for 28.6%. Meanwhile, Huawei, Xiaomi, Apple, Vivo and Oppo, the top five mobile phone manufacturers in China, occupied 91% of the total market share. The proportion of China's second-hand and refurbished mobile phones is similar to that of the Netherlands, and refurbished iPhones have the largest profit and dominate the share of the total number of refurbished phones. In Poland, mobile phones from global manufacturers based in East Asia also dominate the market, e.g., Samsung (30.39%), Huawei (29.14%), LG (12.71%), Xiaomi (5.07%), Lenovo (3.22%), and Sony Ericsson (3.06%) [24].

While mobile phone repair constitutes a business response to an economic opportunity in Europe, complementing the civil society responses of product repair and repair and resale activities of social enterprises (organized in Europe via RREUSE), in China, the story of mobile phone repair is slightly different. In the early stage of the mobile phone industry, after-sales services for mobile phones were implemented by various manufacturers, where mobile phones were collected by the authorized service stations and sent to the central maintenance service station for repair. To deal with the problems of slowness, mobile phone manufacturers launched a rapid maintenance service and set up a rapid response maintenance team in the first level market, to provide after-sales service for users more quickly. With increasing competition in the mobile phone industry, and the emergence of more and more domestic mobile phone manufacturers in particular, the profit level of the entire mobile phone industry is decreasing. Because the quality of mobile phone after-sales services had a great impact on the reputation of the mobile phone brands, and because the construction of a wide range of after-sales service networks required not only substantial funds but also excellent chain management abilities, mobile phone manufacturers gradually outsourced the after-sales service of their phones. For this reason, more and more third-party independent maintenance organizations appeared in China [13,25].

In this sense, to analyze the competitive dynamics, we use Porter's 'diamond' model as the principal framework [26,27]. By incorporating firms' strategy, structure and rivalry into a wider framework made up of demand conditions, factor conditions, government policy, and related and supporting industries, the 'diamond' model is an appropriate model for the analysis of mobile phone repair activities in the context of a circular economy (Figure 1). Each of the five elements is thus viewed as being relevant for the purpose of our article and analysis. Criticisms of Porter's model that it does not give sufficient attention to global interactions and the role of multi-national corporations [28–30] do not pertain to our analysis, which explicitly looks at international supply chains and the decisions of multinational mobile phone producers regarding product design, spare parts production, and repair and remanufacturing. Our analysis is not used to investigate the propositions of Porter's framework (e.g., that rivalry is critically important in pressuring companies to innovate, to cut costs, and to improve quality). The framework is used as an analytical framework, adapted to the analysis of glocal business activities of independent mobile phone repair shops.

Competition/rivalry is a key phenomenon in the mobile phone repair industry. Independent repair shops compete in local markets with one another and with original equipment manufacturer (OEM) repair shops. Although official (authorized/certified) repair stores exist, they are challenged to do a very good job in servicing customers who have to pay relatively higher prices, cannot speak directly to the person doing the repair, have to send the phone by mail and have to wait for a considerable time for the phone to be repaired and returned. The relatively poor customer value can be related to the main interest of global mobile phone manufacturers lying in the business of producing and selling new phones, yet being helped by low entry barriers and traditions of informal repair. Independent mobile phone repair shops emerge, enter the market of repair and earn a living repairing mobile phones by providing relatively superior value propositions that fill financial, technical, and even social gaps for local customers.

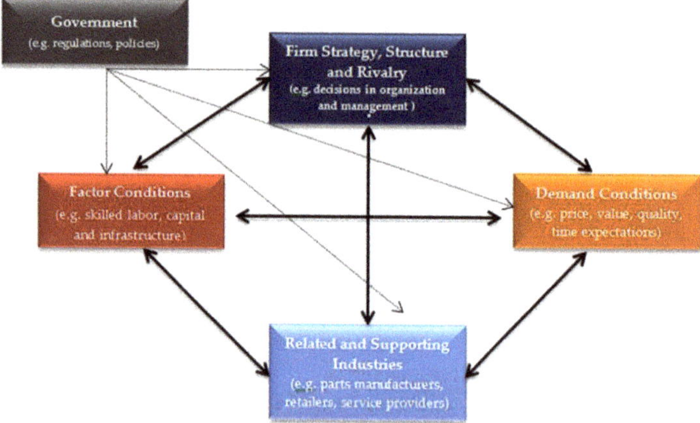

Figure 1. Porter's diamond model. Source: Authors' work.

Government policies also play a role in this. The two-year warranty requirement (Directive 1999/44/EC), extended producer responsibility and waste electrical and electronic equipment (WEEE) legislation in Europe necessitated remanufacturing, reconditioning and/or repair at the manufacturers' side. With such legislation, manufacturers became liable for their products through and beyond their end-of-use life. The activities of repair compete with the options of remanufacturing and recycling. This means that there is not only rivalry between repair companies but also competition between four alternative strategies to reduce end-of-life waste within the context of extended producer responsibility, namely, repairing, reconditioning, remanufacturing and recycling (Figure 2). Original mobile phone manufacturers thus align towards establishing their own official repair services, globally.

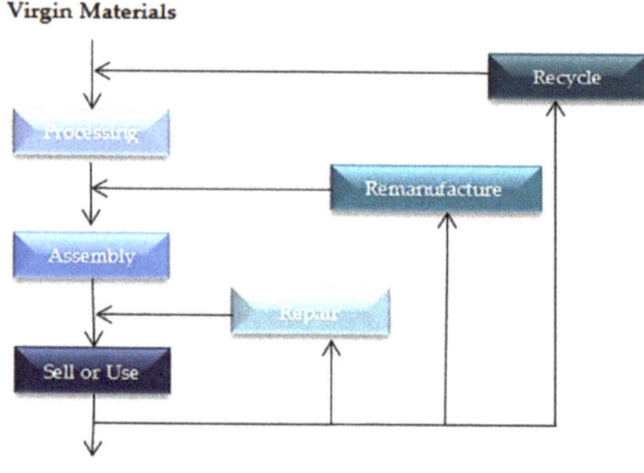

Figure 2. Closed loop design through repair, remanufacturing or recycling. Source: Author's work.

The environmental aspects of mobile phone repair versus remanufacturing, recycling and disposal have been studied by [6,31–36]. The conclusions broadly confirm the statement made by Stahel that the recycling loop (in Figure 2) which uses highly disordered materials, also requires more corrective energy than the remanufacturing loop where the primary shape of the product is preserved [37]. Remanufacturing, on the other hand, in general requires more material, energy and labor skill content (Figure 3) than reconditioning and repair activities [38]. Economic conditions tend to favor the recycling of materials over repair and reuse [37–39], both of which extend the lifetime of products and contribute to the local economy [18]. Nevertheless, this holds less true for mobile phones because of the high prices of new phones from the perspective of mobile phone users.

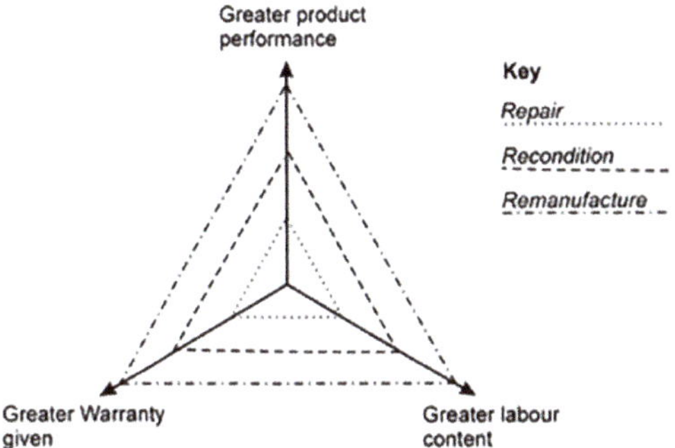

Figure 3. The hierarchy of secondary market production processes. Source: [39,40].

Factor conditions denote specialized factors of production or service, such as conditions related to skilled labor, capital and infrastructure, which give a firm its competitive edge. Independent repair shops benefit from low levels of entry barriers in the form of relevant skills needed in repair activities and financial capital to establish a small or medium size repair shop, and do not rely on sophisticated infrastructure to be able to provide their repair services. Yet, for each of these aspects related to factor conditions, real challenges exist in sustaining the business, due to the advances in mobile phone technologies. Mobile phones by design have become more difficult to repair, and require acquisition of new knowledge and capabilities by repair shops to be able to continue repairing these new devices. In addition to technical skills, social skills and capabilities are becoming more relevant to connect with different generations of mainly local customers. Although initial capital needs to establish a repair shop are relatively low, maintenance of the business necessitates selling additional products and services to create additional streams of revenue ranging from selling accessories and mobile phone call credits, to selling used or new laptops and portable sound systems. Finally, the tools needed to continue repairing new devices, and finding trusted electronics or accessory retailers for parts and additional products, as well as their price dynamics, challenge the repair shops. Shops which can manage to overcome these challenges, through their specialized factor conditions, are more likely to survive and register profits.

For independent repair shops, the presence of related and supporting industries is of critical importance. Six groups of industrial actors make up the mobile phone industry: mobile phone producers, electronics retailers, network service providers, repair services, refurbished second-hand sellers, and accessory producers and retailers (Figure 4).

Electronics producers and retailers, which provide spare parts and components to be used in repair services, form the main supporting industry for repair shops. Activities of refurbished second-hand sellers, such as repair and data removal prior to sales, make used/refurbished phones more common phenomena in the market for customers. Many repair shops also engage in these second-hand repair and sale activities as a side business (Figure 4).

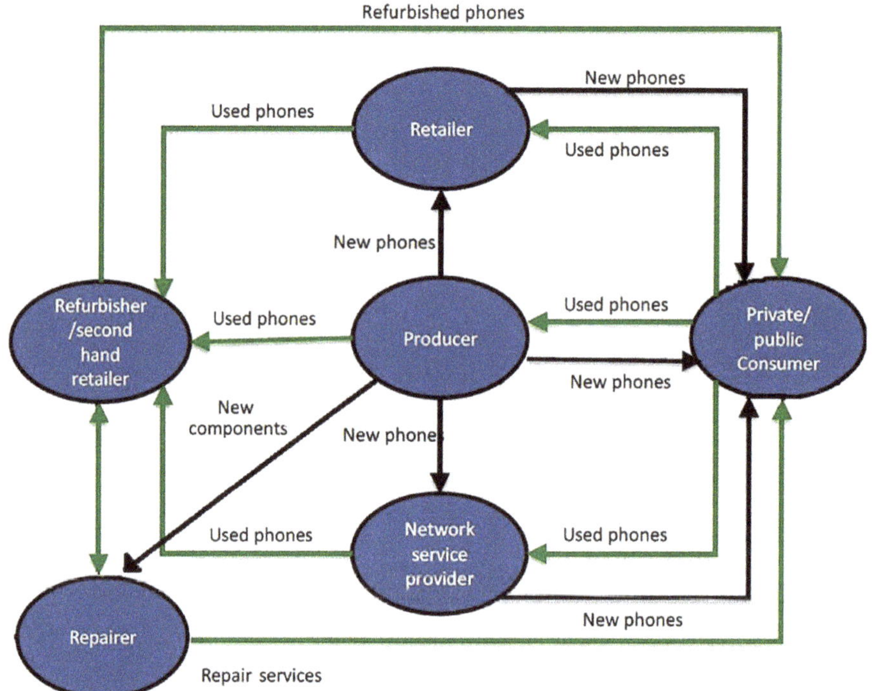

Actors	Linear Activity	Circular Activity
Mobile Phone Producers	-Producing phones (model upgrades and obsolescence)	+Designing phones for durability and ease of repair
	-Producing spare parts	+Provision of spare parts for repair
	-Providing services limited to warranty schemes	+Take-back services
Electronics Retailers	-Providing mobile phones to consumers	+Nudging customers towards greener alternatives
	-Providing mobile phones to businesses	+Entering the second-hand market by selling used mobile phones to refurbishers or repair shops
	-Providing services limited to warranty schemes	+Acting as a collection point for waste electrical and electronic equipment (WEEE)
Network Service Providers	-Selling mobile phones by subscriptions of network services	+Leasing mobile phones and mobile phone upgrades
	-Providing mobile phone upgrades phones by subscriptions of network services	+Buy-back mobile phones and mobile phone upgrades
Mobile Phone Repairers	**By definition circular (maintaining, prolonging the life of product)**	**+Repair by using new parts, +Repair by using used parts, +Trading second-hand mobile phones with warranties to customers and businesses, +Entering the second-hand market by selling used mobile phones to new users or refurbishers and/or remanufacturers, +Acting as a collection point for WEEE, +Selling or sending mobile phones to recyclers.**

Figure 4. Stylized relations among actors in mobile phone industry value chain. Source: [19].

Thanks to their local presence (which allows for direct contacts with clients), skills, ties with global upstream and downstream suppliers, relatively superior price for repair propositions, and the selling of additional accessories and services to customers, independent repair shops are able to create competitive customer value.

3. Data and Method

For this article, we engaged in primary data collection via a questionnaire with closed and open survey questions, and interviews. This questionnaire served the purpose of obtaining factual and qualitative information on a large number of issues with regard to firm strategy, structure and rivalry, factor conditions, demand conditions, relating and supporting industries, the role of governmental interventions, and the dynamics of these repair shops within a circular economy transition context in their glocal business ecosystem. Interviews with participants helped us to more deeply explore specific issues, and a meeting with three repair shops owners in Maastricht helped us to sharpen the question-and-answer categories (and to remove sensitive questions on profits and other issues). Survey questions in English can be found in the Appendix A (questions for the Netherlands, Poland and China were basically the same, but adapted to the local context and language). Table 1 maps sub-sections of our survey to the theoretical background presented in Section 2.

Table 1. Survey development.

Survey Sub-Sections	Corresponding theoretical components and interactions
Establishing the Business	Firm strategy, structure and rivalry
Human Capital Needs	Factor conditions
Part Inventory	Firm strategy, structure and rivalry
Tool and Equipment Inventory	Firm strategy, structure and rivalry
Doing the Business	Firm strategy, structure and rivalry; Related and Supporting Industries
Other Supply Chain Issues	Related and Supporting Industries
Technical Capabilities, Infrastructure and Innovation	Firm strategy, structure and rivalry
Customer Relations	Demand conditions
Threats	Government; Firm strategy, structure and rivalry; Factor conditions; Demand conditions; Related and Supporting Industries
Views on Mobile Phone Manufacturers	Firm strategy, structure and rivalry

The questionnaire was prepared in English, translated into Dutch, Chinese and Polish, and pilot-tested in each country before contextual adaptations are performed. For the case of the Netherlands, it is was sent online to 130 independent mobile phone repair shops after desk research to identify these shops; 17 out of 130 participated in our study (13% response rate). In China, we undertook an on-site inquiry to investigate the data and a total of 76 mobile phone repair shops answered the interview questions. After collation, we received 70 valid questionnaires, which accounted for 40% of the shops contacted. In Poland, both online distribution of the questionnaire and computer-assisted telephone interviews (CATI) were used to reach respondents, resulting in 52 answers from 443 attempts to contact repair shops (an average response rate of 12%).

Our questionnaire with closed and open questions and complementary interviews were the primary methodological instruments used for analyzing the dynamics of the glocal business ecosystem of independent mobile phone repair shops. The response rate for the Netherlands (13%) was rather low because of the difficulties associated with the target population, which was reluctant to participate due to busy work schedules, crowded repair shops during work hours, take-home tasks from work and tiredness after work hours. Since we conducted complementary interviews, on the issue of size, we relied on Guest, Bunce and Johnson, who found that 6 to 12 observations (interviews) are sufficient for a scientific inquiry into a homogenous sample [41]. Our observations are mainly from Limburg and North Brabant provinces of the Netherlands, and apply foremost to these provinces. In China, the research group was divided into four teams, and the survey was conducted in Shanghai, Anhui, Jiangsu and Hubei. Most of the surveys were carried out by students during their summer vacation. A total of 175 mobile phone repair shops were visited, and about 40% (70 repair shops) agreed to accept our interview. Among the feedback survey, there were 18 repair shops in Shanghai (45.7%), 11 in Hefei Province (15.7%), 11 in Jiangsu Province (15.7%), and 16 in Hubei Province (22.8%). In Poland, the questionnaire was first distributed online, sent to 280 mobile phone repair shops, resulting in 12 responses (4% response rate). As the low response rate is a well-known problem in online distributed questionnaire research, to increase the number of responses, on the second stage of the research process, structured interviews were conducted using CATI via a contracted market research company in the Warsaw metropolitan area. After effectively contacting 262 companies sampled as mobile phone repair shops based on the official enterprise register, an additional 40 interviews were conducted (response rate 15%). The results from both stages were analyzed jointly, resulting in 52 observations.

With regard to the response rates above, the probability of capturing a theme within a sample can be approximated by the ratio of the sample size to the target population [42]. A sample of 13% of the target population size indicates a probability value between 0.81 and 0.92 that all themes relevant to our study are captured. In this regard, we judge that our samples for the Netherlands, Poland and China, especially at the provincial levels named above, are representative. In other words, the small size can be expected to not significantly influence our findings and conclusions. While it would have been better to have a larger sample size, the study does not claim strong generalizability, but rather relevant contextual insights that are believed to apply more widely (according to our informers). However, further research is needed for determining the robustness of this argument.

4. Results

Table 2 below demonstrates the findings for selected questions in a comparative way among the Netherlands, Poland and China. In Section 4.1 (The Netherlands), 4.2 (China) and 4.3 (Poland), we analyze the overall results and discuss findings.

Table 2. Selected results from mobile phone repair shops survey.

	Survey Questions	NL (n = 17)	PL (n = 52)	CH (n = 70)
		Establishing the Business		
Q3	3. Are you involved in the business of selling refurbished mobile phones? (yes/no)	47.10% (yes)	23.10% (yes)	15.71% (yes)
Q5	5. How long did it take you to establish your mobile repair shop (from idea to becoming operational)? in months:	2 to 4 years	In a few months	At least 6 months
Q6	6. Before starting an official business, did you repair mobile phones on an informal basis? (yes/no)	29.40% (yes)	15.40% (yes)	65.70% (yes)
Q7	7. What business or other activity (education, unemployed etc.) were you involved in before you established a repair company?	Educational background in IT or IT-related fields. Previous areas of business vary, yet all related to services sector, such as construction services, cooling and air-conditioning services, logistics, car repair, insurance sales, outlet sales, sports and gym management	Very diverse professional experience, such as production work, customer care, advertisement, or watchmaker. Also, education varies from general secondary school to higher education in IT or chemistry	Most of the respondents worked in IT related occupations. A few of the respondents were engaged in other business, such as salesmen and attendants.

Table 2. *Cont.*

	Survey Questions	NL (n = 17)	PL (n = 52)	CH (n = 70)
		Establishing the Business		
Q11	11. Did competition increased in the last year?	76.50% (yes)	50% (yes)	68.57% (yes)
		Part Inventory		
Q17	4. What percentage of the parts is new? ... % of the parts are new	95%	34.60%	68.60%
		Doing the Business		
Q26	2. Do you give a warranty? (yes/no)	100%	100%	74.29%
		Other Supply Chain Issues		
Q31	1. Do you accept phones by postal mail? yes/no	94.10% (yes)	82.70% (yes)	51.43% (yes)
Q32	2. Do you have contact with remanufacturers? yes/no	23.50% (yes)	26.90% (yes)	24.29% (yes)
Q33	3. Do remanufacturers collect phones from your store? yes/no	0% (yes)	15.36% (yes)	42.86%(yes)
Q34	4. Do you have contact with recyclers? yes/no	64.70% (yes)	34.60%(yes)	58.57%(yes)
Q36	6. Do you receive payments for giving phones from recyclers? yes/no	11.70%(yes)	7.68%(yes)	61.43%(yes)
Q37	7. Do you receive or collect phones beyond repair? yes/no	41.20%(yes)	67.30%(yes)	50%(yes)
		Customers		
Q39	1. Do you buy phones from your customers for repair and resale? (yes/no)	64.70%(yes)	42.30%(yes)	24.29%(yes)
		Threats		
Q42	1. What are the main threats for your repair business?			
	Informal repair shops	68.80%	44.20%	37.50%
	Rising prices of parts	37.50%	32.70%	62.86%
	Falling prices of mobile phones	43.80%	38.50%	61.43%
	Replacement plans	12.50%	0.00%	15.71%
	New repair shops	62.50%	38.50%	42.86%
	Provisions on employment Conditions	6.30%	13.50%	5.36%
	Provisions on maintenance product safety	0.00%	7.70%	12.86%
	Leasing	12.50%	11.50%	8.57%
	Provisions on operation and maintenance business	43.80%	15.40%	20.00%
	Rising maintenance tools and equipment prices	6.30%	13.50%	38.57%
	Official shops	43.80%	19.20%	50.00%
Q43	2. Have phones become difficult to repair? (yes/no)	58.80% (yes)	69.20% (yes)	71.42% (yes)
		Your Views		
Q44	1. Could manufacturers do more to avoid the need for repair?	35% (yes)	51.9% (yes)	50% (yes)

4.1. The Netherlands

4.1.1. Informality and Transitioning to Official Repair Business

In the Netherlands, 29.4% of respondents indicated that they repaired phones in an informal way before they started their business. In line with this, provisions on the operation and maintenance of business are reported as an important business threat by 43.8%. Before getting into mobile phone repair activity, half of the interviewees report an educational background in IT or an IT-related field. Previous areas of business vary, yet all related to the service sector, such as construction services, cooling and air-conditioning services, logistics, car repair, insurance sales, outlet sales, and sports and gym management. Interestingly, many migrants are active in the mobile phone repair sector both as owners and workers. One reason for this is that migrant owners prefer to recruit migrant workers. Informal activities usually take 2 to 4 years in transitioning to an official shop.

4.1.2. Motivation Behind Establishing a Repair Shop and Profitability

In the Netherlands, business experience of former shop owners is deemed not so important. We observe that high profits and low risk still play an important role, if not a crucial one, in the Dutch context, and, in general, 6 months is the average duration to reach profitability. This is highly likely to be due to the mobile phone penetration rate and the higher average price of the devices owned in the Netherlands.

4.1.3. Threats in the Business Ecosystem

In the Netherlands, informal repair shops and new repair shops are the biggest threats (68.8% and 62.5% respectively). Increasing prices of spare parts and falling prices of mobile phones also affects the Dutch context. However, rising prices of maintenance tools and equipment and provisions on employment conditions do not threaten the repair shops (6.3%). Official repair shops and provisions on the operation and maintenance of the business pose considerable business threats according to the respondents (each 43.80%) (Figure 5).

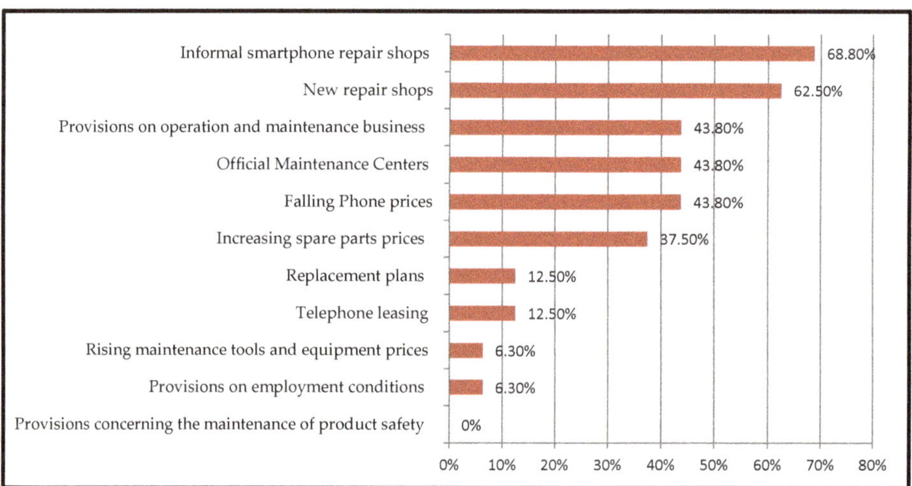

Figure 5. Business threats to mobile phone repair shops in the Netherlands.

Interviewees in the Netherlands mentioned defensive actions of mobile phone manufacturers as a business threat. One interviewee indicated that "Certain brands do not make available parts for non-official repairers, which makes it difficult to guarantee quality". Mobile phone manufacturers usually work with their own official repair services, and they also engage into certification of other shops to authorize them. For non-authorized shops, these authorized/certified repair shops are a business threat. One interviewee indicated that once repair shops get authorization certificates, they deprecate unauthorized/non-certified repair stores: "We are often blackened by certain companies that say we are very bad, and these companies have a certificate". However, the defaming of non-certified repair shops is not completely without grounds. An interviewee admits that "Some independent repair shops create a very bad image for the entire independent repair shops by using very low quality and cheap spare parts" and this results in misinformation and, thus, an unfair competitive environment. Competition is skewed to companies with advertising power. An interviewee points out that "Official stores use online advertisement AdWords; they pay to pop up first in Google Search". The business environment for mobile phone repair business becomes more challenging as a greater number of informal and illegal repair activities enter in the market: "There are more and more homeworkers … illegal businesses". Yet not all threats are external; repair shops are challenged also by internal weaknesses in keeping up with the new knowledge: "Repairing becomes more difficult and the knowledge is not always there … maybe also because of my age". In the Netherlands, 58.8% of respondents pointed out that mobile phones have become more difficult to repair.

Considering business threats, as expected, in the last year (2017) competition increased for 76.5% of respondents in the Netherlands: "Unfortunately, there are too many shops that offer repair for a very low price".

4.1.4. Supplier Relations

In the Dutch context, 59% of the repair shops change their suppliers quickly for cost reasons (47.1% do not). Formal and informal contracts are equally used. The origin of main suppliers is China. Direct supply accounts for 76.5% of mobile phone repair shops in the Netherlands. As well as the low price, the variety and reliability of supply are also important determinants for the choice of suppliers. Most use China as a preferred supplier. Some rely on intermediaries: "It was easier at the beginning but now the quality parts are hard to find if you have to buy them yourself. The big buyers in the Netherlands know their way and therefore also find good parts at a reasonably attractive price." Interviewees also indicated that components are often not available in Europe, and that even the original parts are available in China. "Big buyers are mainly located in the Netherlands, if we cannot find what we are looking for from them, we contact big buyers in Germany or in the UK".

4.1.5. Customer Relations

Social interactions with customers play an important role in establishing a mobile repair shop in the Netherlands (76%), and keeping up with these social relations is deemed as a key activity. Interestingly, 66.7% of the shops have decided not to use such parts to maintain the trust between them and their customers, and to provide best customer satisfaction. Very much in line with this, 58.8% offer paid and unpaid services outside the repair of components in the Netherlands. The most common service provided to customers is related to software (e.g., software recovery, support, reset and replace), followed by data related services (e.g., data backup, transfer, recovery, memory reset). Financially, these services form an important part of the revenues of only 31.3% of firms. Reductions, discounts, or even no billing in some cases for such services are used to maintain good social interactions with customers, sustain customer satisfaction and enhance customer loyalty. Repair shops also extend their services to logistics companies as contact points, e.g., for DHL and Western Union. In line with this, 94.1% of respondents accept phones by postal mail.

4.1.6. Circularity Oriented Economic Relations

Only 23.5% of the surveyed repair companies indicated having contacts with remanufacturers. Remanufactures do not collect phones from independent repair shops. A total of 64.7% of the repair shops is in contact with recyclers, but only 17.6% send or sell to recyclers, where only 11.7% of the repair shops are being paid by these recyclers for these sent phones. In addition, 41.2% also collect mobile phones beyond repair, 47.1% sell refurbished phones, especially after 2015, and 52.9% are interested in using parts from used phones more often. A total of 64.7% of respondents buy used phones to sell later.

Contact with recyclers, as well as selling refurbished phones, are established practices. According to Deloitte (2017), 12% of mobile phones in the Netherlands are either reused or refurbished. An interviewee indicated that "Refurbished market is rising enormously. Many customers ask for used or refurbished devices. Due to the compulsory Bureau Krediet Registratie (BKR) registration for phones over 250 Euros bought with credit, many people buy a separate device instead of taking a subscription with registration". Environmental reasons, customer demand, and cost reasons are the most important motivations behind using components from used phones; supply constraints for new parts is the least important factor in Dutch context. On this, interviewees indicate that "Used original parts do not mean that they are worse. Some copy parts that are new are less good as a used original. So used parts are not bad from experience and it is also much better for the environment".

Using used parts necessitates testing and careful disassembly: "As long as every part is tested, that's well, nothing is wrong. But then you have to start looking again that these parts do not become more expensive than the new parts". Another interviewee notes that parts from used devices are not tested properly. "Often these parts come from water damaged devices. It is a nice idea to get parts from used equipment, but it requires paying attention to disassembly which is not common if you buy used spare parts from somewhere else".

New parts are used by 95% of respondents. Paying for a used phone is different than paying for a used part to be utilised in repair. Customers do not prefer repairs with used parts and prefer new parts. Repair shops provide a 100% warranty in the Dutch context, which is in line with the finding that provisions concerning the maintenance of product safety are reported as a zero threat. All respondents request a supplier guarantee in the Netherlands.

4.1.7. Planned Obsolescence and the Future of the Repair Business in the Netherlands

Many interviews said that mobile phones are designed to have a short life. According to one independent repair shop owner "Manufacturers' initial aim is the sale of a new device every two years, making the device weaker so that the device does not last longer than 2 years. Yet they also earn on the parts". Another interviewee reflects on the technological advances at the side of manufacturers and consumer attitudes towards newness: "We are a disposable industry and that is not only because of less quality but also because of design, innovation, new and better techniques, developments (e.g., stronger glass (possibility is already available with sapphire glass) … but certainly, also because of the urge of people to own something new and be able to afford it". Planned obsolescence is a complex phenomenon involving hardware-based weaknesses, software-based update requirements, as well as new design features which make an older yet fully functioning model undesired in the eye of the customers. Although measures are taken for hardware-based weaknesses (e.g., waterproof devices, stronger glass in screens), software-based obsolescence still remains as an important part of the planned obsolescence debate. Yet, in the Netherlands, an increase in the demand for mobile phone repair is also foreseen by repair shop owners due to increasing awareness (e.g., technical, economic, legal) of the demand side: "It is becoming increasingly known to people that a device can still be repaired. It is also becoming increasingly known that there is a difference in the quality of repairs/parts", and due to government interventions, changes in law relating to subscriptions, and "especially, due to changes in the law and change of subscription structure".

4.2. China

4.2.1. Informality and Transitioning to Official Repair Business

In China ($n = 70$), 65.7% of respondents reported that they repaired mobile phones on an informal basis before they officially started business. Before getting into mobile phone repair activity, most of the respondents worked in IT-related occupations. A few of the respondents were engaged in other businesses, such as salesmen and attendants. Compared with the Netherlands, mobile phone repair shops in China usually do not involve immigrant owners. It usually took at least half a year for informal activities to transition into an official shop in China.

4.2.2. Motivation Behind Establishing a Repair Shop and Profitability

Prospects of high profits (47.1%), low risk (41.4%) and previous experience (42.9%) are considered very important drivers in establishing their shops. Within one year after opening, 50% report that they reach profitability. The large number of mobile phone users and the fast updating of smart mobile phones in China are positive contributing factors.

4.2.3. Threats in the Business Ecosystem

The main threat in the business ecosystem is financial: rising prices of parts (62.86%) and falling prices of mobile phones (61.43%). Next to this, official repair centers (50%) and new repair shops are said to pose a serious threat (42.86%) for the independent repair shops. Rising prices of maintenance tools and equipment (38.57%) are a bigger threat than informal repair shops (37.5%) in China. Provisions for operating and maintenance business (20%), replacement plans (15.71%), provisions on maintenance product safety (12.86%) are less pressing, while phone leasing (8.57%) and provisions on employment conditions (5.36%) form the least significant threats according to the respondents. Non-repair activities are engaged in by 61.43% of respondents for additional revenue. In the context of China, competition in the mobile phone repair market has increased for 68.57% of respondents (Figure 6).

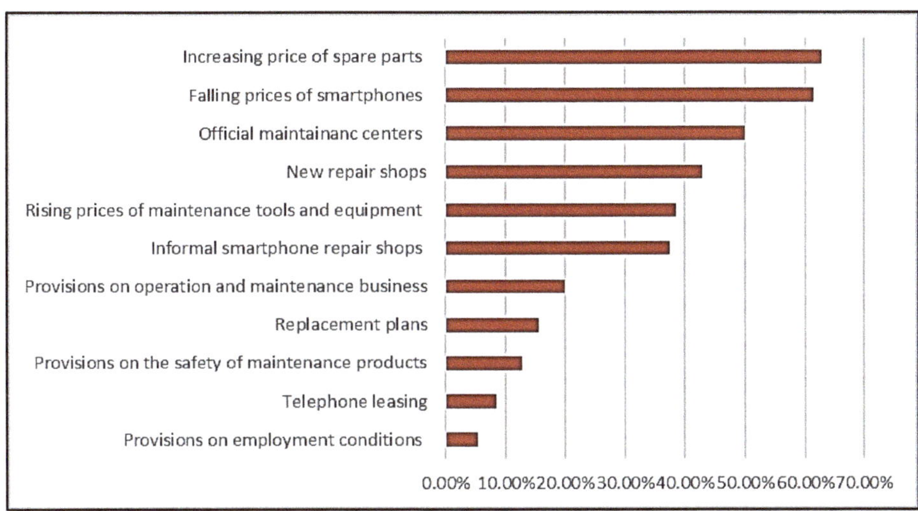

Figure 6. Business threats to mobile phone repair shops in China.

In interviews, Chinese interviewees noted that the increasing renewal rate of mobile phones is threatening their business. Many people will renew their mobile phone within 2–5 years. In the first year, mobile phones are under the quality warranty period. During this warranty period, consumers will go to official repair centers for fixing their phone. After around 2 years, many people would prefer to replace a new mobile phone instead of repairing it. Several interviewees also said "Many people worry that private repair shops cannot completely repair their mobile phones or complain that the maintenance cost is too high. In fact, after the warranty period of the mobile phone, the repair cost in the private repair shop is lower than in the official repair center."

Furthermore, with the frequent updating of mobile phones, mobile phone repair centers are also facing threats on the technology side. Our survey revealed that 71.42% of our investigated shopkeepers in China think that mobile phones are becoming more and more difficult to repair. Their feedback reflects that some issues of mobile phones are not easy to handle.

4.2.4. Supplier Relations

A total of 97.14% of respondents sources parts from China (home country); 68.57% of the parts are new (31.43% are used) and 84.29% require suppliers to provide warranty on these parts. Original new parts are sold by 77.14% and 50% quickly replace the parts supplier for cost reasons. Shops have formal contracts (30%) and informal relations (34.29%), and 35.71% have both. These findings indicate the knowledge and cost-based constraints of repair shops in China.

Regarding quality and price, it is said that "mobile phone accessories in China are adequate and inexpensive. The parts made in China can cope with normal repair work. Usually there is no need to use foreign parts". The mobile phone industry in China has made rapid progress. More than 70% of the world's mobile phones are produced in China and eight of the top ten mobile phone sales brands in the world are from China, making it the first option for mobile phone parts supplies.

4.2.5. Customer Relations

Phones are accepted by postal mail by 51.43% of respondents. A total of 75.71% repairs all types of mobile phones, and 31.43% even repair illegally made bandit phones. Warranty for services are provided by 74.29% of respondents, and 42.86% provide fee-based services other than repairs. For 65.71%, these services are an important source of their revenues, while 35.71% plans to provide fee-based services in the future. These findings indicate the increasing need of shops in being able to provide additional services surrounding repair activities in China.

It is said that "The online business system for mobile phone repairing in China is becoming popular and mature. Customers do not need to show up in our shops. Usually, we can receive customers' orders on Taobao. Then customers will send us their phones and we send phones back to them after repair work finished". This kind of online business mode offers convenience to both sides. The convenience is based on mature online communication and monitoring business system, and trust between the customer and repair shops. "The online order sometimes can be a big part of our business, this undoubtedly increases our income", some interviewees said.

4.2.6. Circularity Oriented Economic Relations

In China, 24.29% of the repair shops are in touch with the refurbished manufacturers. Refurbished manufacturers source phones from repair shops (42.86%). While 24.29% buys a customer's unwanted phone for repair and resale, only 15.71% of the repair shops sell refurbished phones. A total of 18.57% of respondents is more interested in using used mobile phone parts than in using original parts yet are respondents worry that consumers do not want such parts (61.43%). Warranty and legal requirements (34.29%) are the most important reasons why shops do not prefer using used mobile phone parts. Some interviewees stated that using used parts would save a lot of costs, but the quality of old parts is difficult to guarantee. Since the new parts are not expensive in China, the mobile phone repair shop owner usually uses new parts rather than old ones.

Regarding recycling, 58.57% of respondents has contact with recyclers, 61.43% also sells phones to recyclers, and 50% recycles phones other than repairing them. Many of the recycled used phones go to professional "online trading system for second hand phones", such as Taobao and Yiji. The used phones would be tested and given a recycled price. These findings indicate the increasing maturity of circular economic relations of mobile phone repair shops and the subsequent phone reuse and recycling business in China.

4.2.7. Planned Obsolescence and the Future of the Repair Business in China

In China, 50% of the respondents think that manufacturers can do more to avoid maintenance. According to the participants, manufactures do not want to extend the use of mobile phones for more than 2 years (planned obsolescence). A total of 42.86% indicate software requirements and updates require that a new phone will be needed in any case. On the other hand, 55.36% still argue that if more people buy mobile phones, this will lead to more repair and maintenance demand; for 28.57%, more mobile phones sold would not make any difference to the demand for repair.

With regard to the future of the mobile phone repair industry in China, our respondents generally believe that the industry is in a period of decline. The main reason is that our respondents believe the profit margin of mobile phone repair is decreasing with the continuous upgrading of mobile phones, since the technical requirements for fixing high quality mobile phones, such as those of Apple, are increasing.

4.3. *Poland*

4.3.1. Informality and Transitioning to Official Repair Business

In Poland, before getting into mobile phone repair activity, interviewees had very diverse professional experience, such as production work, customer care, advertisement, or watchmaker. Also, education varies from general secondary school to higher education in IT or chemistry. Only 28.8% indicated that they worked in similar business—repairing phones or computers—before they opened their own repair shops, and 15.4% said that they repaired phones in an informal way before they started their business. Transitioning to an official shop can be done in a few months' time.

4.3.2. Motivation Behind Establishing a Repair Shop and Profitability

In Poland, the most important motivation to open a repair shop is the expectation of high profits and the former business experience (important or very important for 86.5% of respondents). Almost equally important is the preference to be the owner (not to work for somebody else) and opportunity to work with people (84.6%). As many as 78.8% of respondents point out that they simply like the repair work. Furthermore, 61.5% mentioned low risk as an important factor. This indicates that in Poland, risk connected to the establishing of the repair shop is not considered to be low. At the same time, most of the respondents in Poland claimed that the company started to generate some profits as early as after one month (44.2%), while 23% indicated a period between 2 and 6 months. Usually, after taking the decision, it takes only 2–3 months to open the shop.

4.3.3. Threats in the Business Ecosystem

In Poland, generally, respondents are less likely to see changes in the environment as a threat to their business than in the Netherlands and China. Similar to the Netherlands, informal repair shops (44.2%), new repair shops and falling prices of mobile phones (both 38.5%), and increasing prices of spare parts are the biggest threats (32.7%). We can see that the same sources of threat are repeated between countries as being most relevant (Figure 7).

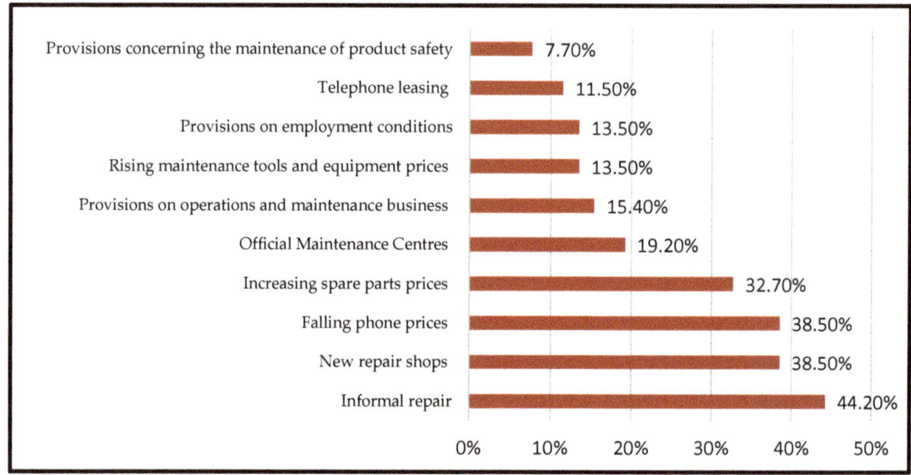

Figure 7. Business threats to mobile phone repair shops in Poland.

Additionally, respondents mentioned product change as a source of threat: progressing miniaturization already made some repairs very difficult and up to some point, may render it impossible.

The development of "non-repairable" mobile phones is seen by some as a purposeful action of the producers. Indeed, 69.2% respondents in Poland point out phones are currently more difficult to repair then they used to be. Moreover, 50% believes that competition increased in the last year, while 36.5% does not see the changes in this regard.

4.3.4. Supplier Relations

Only 23% (in comparison with 59% in Dutch context) of Polish repair shop owners change their suppliers quickly for cost reasons; 32.7% of owners base these relations on formal contracts, 48.1% rely on informal contracts, while 19.2% develop both formal and informal contracts. Almost all cooperate with suppliers in Poland (88.5%), while a considerable number also buy spare parts or tools from China (40%). A few also indicated countries such as Germany or England. Those who obtain supplies from China point out their low prices (76.2%) and wide assortment (61.9%). When parts are not available in Poland, they would buy directly from Chinese suppliers, in particular.

4.3.5. Customer Relations

Opportunity to work with people is highly valued by repair shop owners in Poland. To increase client satisfaction and, possibly, to generate the additional source of income, it is not uncommon for additional services are offered besides repair: 19.2% of respondents offer such services, such as parcel pick-up, short-term lease of electric equipment, or phone/computer configuration. Only for 5.8% of respondents did these activities constitute a significant source of income. The acceptance of phones sent by mail is widespread: 82.7% of respondents accept them.

4.3.6. Circularity Oriented Economic Relations

Similarly to the Netherlands, 26.9% of respondents pointed out that they have regular cooperation with remanufacturers. From this group, 57.1% have arranged the regular collection of phones from their repair shops. Only 34.6% of companies are in regular contact with recyclers, and from this group, 22.2% receive payment from recyclers for the phones which they pass to them. As much as 67.3% of respondents also collect mobile phones beyond repair, while 42.3% buy used phones to sell later and 23.1% sells refurbished phones.

In Poland, 65.4% of respondents use used parts in their repairs, a share that is considerably above that in China (31.4%) and the Netherlands (5%). They point out the following reasons: limited access to new parts (70.6%), lower costs (38.2%), and customers' expectations (29.4%).

Environmental protection is less important (11.8%). Respondents point out that used components have higher quality—they claim that "it's always better to get the part from the original phone, rather than cheap replacement".

All repair shops provide a warranty, usually for three months (63.5%) or six months (25%).

4.3.7. Planned Obsolescence and the Future of the Repair Business in Poland

More than half of the respondents in Poland (51.9%) believe that producers could do more to avoid the mobile phone repairs; in other words, they suspect producers apply a planned obsolescence strategy. In their opinion, producers do this because they want their customers to buy a new device after the old one is broken. Some of the research participants believe that the new mobile phone models are also designed to hinder the repair; currently, it may be even difficult to open the mobile phone to start the repair process.

The low level of perceived threats to the business signals the general optimism of Polish repair shop owners. This may result from their evaluation of the impact of rising sales of mobile phones: 73.5% of respondents believe that more mobile phones on the market will lead to more repairs. However, they also stress that it depends on the value of the mobile phone: the more expensive phones are repaired, while the cheaper phones are replaced.

5. Discussions and Conclusions

In this article, we examined the glocal competitive dynamics in the mobile phone repair sector in the Netherlands, China, and Poland. In doing so, the article contributes to the literature on the circular economy by offering an in-depth study of independent shops specializing in mobile phone repair in different local contexts, which emerged in a self-organized way, without special government innovation programmes and support schemes. The analysis was done with the help of survey questions with closed answer categories and open-ended questions, focus group meetings, and interviews of independent mobile phone repair shops operating in three different contexts.

5.1. Price Squeeze

Facilitating factors for the emergence of their businesses are: *the high prices of mobile phones, official repairs and original spare parts*; technical skills that were easy to learn; relatively quick time for repairs; and the possibility for direct contact with customers in shops. In all three countries, we observe a dynamic interaction between factor conditions and demand conditions in building up competitiveness.

For Chinese repair companies, the biggest business threats stem from rising prices of parts (62.86%) and falling prices of mobile phones (61.43%), followed by rising prices of maintenance tools and equipment (38.57%). The least significant threats include telephone leasing (8.57%) and provisions on employment conditions (5.36%). A negative factor for customers in China is the limited warranty that is generally available from Chinese repair shops, something which is not the case for the repair shops in the Netherlands and Poland because of EU regulations.

5.2. Increasing Technical Complexity

Our results and analysis indicate that changing factor conditions of technical skills play a crucial role for independent mobile phone repair shops. With advances in the mobile phone industry and increasing difficulty in reparability, in order to survive, independent repair shops develop skill-based strategies to cope with new knowledge needs. Keeping up with the new information and knowledge needs and requirements is a challenging activity for independent repair shops in the ecosystem.

Reparability is highly dependent on the choices of mobile phone producers in terms of durability, design, and provision of spare parts associated with their products. This constitutes a dynamic interaction between factor conditions and related and supporting industries, which has a direct impact on repair shops' strategy, structure and rivalry. People opt for mobile phone repair chiefly for financial and technical reasons of not being to solve the issues of their phones themselves.

5.3. *Increasing Competition*

Competition is increasing in the contexts of all three countries, with the Chinese repair shops more negatively affected by price dynamics than the shops in the Netherlands and Poland. To remain competitive, repair shops offer additional services in the form of selling accessories, mobile phone call credits, software recovery (e.g., support, reset and replace), followed by data-related services (e.g., data backup, transfer, recovery, memory reset).

5.4. *Circularity Relations*

Refurbished and repaired phones are common in all contexts, but in China and Poland repair shops have more intensive financial relations with recyclers than in the Netherlands. In all countries, refurbishment and reuse are common activities, in contrast to remanufacturing-related activities, which are still in an emerging stage. For mobile phone companies, remanufacturing is not a priority and neither is sourcing it out to third party suppliers. Remanufacturing and recycling firms are important for independent repair shops, i.e., to buy phones beyond repair from shops to recycle, or providing remanufactured phones to be sold. Nonetheless, such market transactions depend on the choices of mobile phone manufacturers and government regulations.

Relevant policies are Extended Producer Responsibility requirements that define the conditions in which an electronic product can or should be collected, recycled and recovered (e.g., WEEE (2002/96/EC); battery directives (2006/66/EC) in EU); regulatory reuse requirements; and warranty schemes for second-life products. This legislation regulates the activities of these supporting and related (remanufacturing and recycling) industries in business-to-business transactions, and, thus, their interaction with independent repair shops.

5.5. *Policy Recommendations and Future Research Directions*

Our case analyses reveal that the future of local independent repair shops dynamically depends on the circular choices (durability, design, and reparability) of global actors (e.g., mobile phone manufacturers) as key players, and outer-circle remanufacturing and recycling industries of a circular economy. While there are local economy and (digital) platform elements (of on-line repair manuals), mobile phone repair is not purely local because of the non-local availability of product components and materials. Profit levels are decreasing for independent repair shops. However, whether the viability of the sector as a whole is undermined by technical difficulties of repair that stem from design choices of global manufacturers using linear economy business models could be investigated further. Additionally, future research directions on local mobile phone repair may concentrate on the dynamics of global manufacturers' choices in using durable and easy to repair product designs, availability of spare parts for longer periods, and the dynamics of access to repair service documentation and software by third party shops and individuals.

The case of mobile phone repair shows possibilities for business-based circularity action at the local level but also demonstrates the importance of the supra-level (e.g., component suppliers and OEMs in other parts of the world, recyclers, national governments and the EC as the governing body of the European Union). To help mobile phone repair, EU and Chinese legislations could stipulate demands for reparability and information access. A complication is that the business of independent mobile phone repair is poorly organized, making it hard for repairers to argue their case to national or urban-level policy makers. The possibility to directly discuss issues of repair with mobile phone repairers and obtain additional services is greatly appreciated by customers, and a key reason behind the existence of mobile phone repair businesses. Such activities are undertaken as part of global value chains and networks, in which the ties with remanufacturing and recycling firms are, so far, relatively weak. A notable development in this respect is the creation of the fair phone, which is innovatively using the demand side of its supply chain to drive sustainability [43].

Finally, it should be noted that product repair by independent shops is a business response to an economic opportunity. It is not based on shared value creation or sustainability thinking, which are motivators for repair cafés (engaged in the repair of household appliances and other mobile phones) and the (broader) movement of makers, modifiers and fixers [44,45]. Thus, such centers and cafés, and the dynamics of local socio-technical skill formation and non-market-based solutions in mobile phone repair, could be interesting topics of further research.

Author Contributions: Conceptualization, S.T. and R.K.; methodology, R.K. and S.T.; validation, S.T., B.H., and A.S.; formal analysis, S.T., B.H., A.S., R.K.; investigation, S.T., B.H., A.S.; resources, S.T., R.K; data curation, S.T., B.H., A.S., R.K.; writing—original draft preparation, S.T.; writing—review and editing, S.T., B.H., A.S., R.K.; visualization, S.T., B.H., A.S.; supervision, R.K.; project administration, S.T.; funding acquisition, R.K". All authors have contributed substantially to the work reported.

Funding: This research was funded by Nederlandse Organisatie voor Wetenschappelijk Onderzoek (NWO) for Dutch research (467-14-154) SINCERE (Sino-European Circular Economy and Resource Efficiency) project, grants from the National Natural Science Foundation of China (Nos. 71461137008; 71690241; 71403170), and the UK Economic and Social Science Research Council (ES/L015838/1).

Acknowledgments: We would like to acknowledge the implementation support of the field research teams, Madelon Boersma for the Netherlands team, and the students of University of Shanghai for Science and Technology.

Conflicts of Interest: The authors declare no conflict of interest. The funders had no role in the design of the study; in the collection, analyses, or interpretation of data; in the writing of the manuscript, or in the decision to publish the results.

Appendix A

Mobile Phone Repair Survey Questions

Establishing the Business (11)

Q1 1. What are the activities your business is involved in?
Q2 2. What is the share of revenue from repair activities in all activities? (%)
Q3 3. Are you involved in the business of selling refurbished mobile phones?
Q4 4. When did you enter the mobile phone repair business? (The year in which you became operational)
Q5 5. How long did it take you to establish your mobile repair shop (from idea to becoming operational)? in months:
Q6 6. Before starting an official business, did you repair mobile phones on an informal basis? (yes/no)
Q7 7. What business or other activity (education, unemployed etc.) were you involved in before you established a repair company?
Q8 8. What role did the following play in establishing a mobile repair shop?
The prospects of high profits
Low Risks
Previous Business Experience as shop owner
I like repair work
I like being a shop owner
Social interactions with customers
Other:
Q9 9. How quickly was the repair of mobile phones profitable? Please only answer for activities related to the repair of mobile phones. In .... Months / in .... ..years
Q10 10. How many repair shops are there in your area (within 1 km^2)?
Q11 11. Did competition increase or decrease in the last year?

Human Capital Needs (2)

Q12 1. How many employees do you have?
Q13 2. On average, how many telephones could an employee repair per day? (a range can be given when people desire to do this, for example: 5 to 10 phones)

Part Inventory (9)

Q14 1. From which countries do you purchase the parts (components)?
Q15 2. If you purchase parts from China, why?
Q16 3. Do you source parts via trade intermediaries? If so, in which country they are based?
Q17 4. What percentage of the parts is new? ... % of the parts are new
Q18 5. Do you ask for a warranty for the parts?
Q19 6. Are you interested in using components from used phones more often?
Q20 6.1 What are the reasons why you are more interested in using components from used phones?
Q21 7. What is the most important reason for NOT using parts from used phones more often?
It is easy to get original parts
Legal requirements for warranty
Fears that consumers do not want this
Absence of warranty
Other reason (e.g., negative image)
Q22 8. Do you quickly change part suppliers for cost reasons?
Q23 9. Is your relationship with suppliers based on formal contracts or informal relations?

Tool and Equipment Inventory (1)

Q24 1. What is the origin of the tools to repair mobile phones? Please state the most important country e.g., China, other country (to be named)

Doing the Business (5)

Q25 1. Do you repair all types of phones? Yes/no. If no, what phones do you not repair?
Q26 2. Do you give a warranty? Yes/no. If no, what warranty do you give?
Q27 3. Do you repair illegally produced phones?
Q28 4. Do you offer payable services beyond the repair of components?
Q29 5. Do such services constitute an important source of income for you?
Q30 * Do you intend to offer payable services in the future?

Other Supply Chain Issues (7)

Q31 1. Do you accept phones by postal mail? yes/no
Q32 2. Do you have contact with remanufacturers? yes/no
Q33 3. Do remanufacturers collect phones from your store? yes/no
Q34 4. Do you have contact with recyclers? yes/no
Q35 5. Do recyclers collect phones from your store? yes/no
Q36 6. Do you receive payments for giving phones from recyclers? yes/no
Q37 7. Do you receive or collect phones beyond repair? yes/no

Technical Capabilities, Infrastructure and Innovation (1)

Q38 1. Have you bought new innovative tools and equipment in the last year? If so, what was new?

Customers (3)

Q39 1. Do you buy phones from your customers for repair and resale?
Q40 2. What percentage of your sales is from phones which are repaired and resold? ... %
Q41 3. Do you foresee an increase or decrease in telephone repairs in general? Why?

Threats (2)

Q42 1. What are the main threats for your repair business?
Q43 2. Have phones become easy to repair?

Your Views (2)

Q44 1. Could manufacturers do more to avoid the need for repair?
Q45 2. If more people would buy the phone, would this lead to more repairs or less repairs, and why?

References

1. European Commission. Directive 2002/96/EC of the European Parliament and of the Council of 27 January 2003 on Waste Electrical and Electronic Equipment (WEEE). *Off. J. Eur. Union* **2003**, *37*, 24.
2. European Commission. Directive 2008/98/EC of the European Parliament and of the Council of 19 November 2008 on Waste and Repealing Certain Directives (Waste Framework Directive). 2008. Available online: https://eur-lex.europa.eu/legal-content/EN/TXT/?uri=celex%3A32008L0098 (accessed on 15 December 2018).
3. European Commission. Circular Economy Strategy-Closing the Loop—An EU Action Plan for the Circular Economy. 2015. Available online: http://ec.europa.eu/environment/circulareconomy/index_en.htm (accessed on 15 December 2018).
4. Ellen MacArthur Foundation. Ellen MacArthur Foundation—Rethink the Future. 2011. Available online: http://www.ellenmacarthurfoundation.org/circular-economy (accessed on 15 December 2018).
5. Quariguasi, F.J.; Bloemhof, J. An analysis of the Eco-Efficiency of remanufactured personal computers and mobile phones. *Prod. Oper. Manag.* **2012**, *21*, 101–114. [CrossRef]
6. Velmurugan Manivannan, S. Environmental and health aspects of mobile phone production and use: Suggestions for innovation and policy. *Environ. Innov. Soc. Transit.* **2016**, *21*, 69–79. [CrossRef]
7. Ramani, S. Groundhog Day or Tipping Point for the Circular Economy? 2018. Available online: https://www.merit.unu.edu/groundhog-day-or-tipping-point-for-the-circular-economy/ (accessed on 15 December 2018).
8. Statista. Number of Mobile Phone Users Worldwide from 2013 to 2019 (In Billions). 2018. Available online: https://www.statista.com/statistics/274774/forecast-of-mobile-phone-users-worldwide/ (accessed on 15 December 2018).
9. Statista. Number of Smartphones Sold to end Users Worldwide from 2007 to 2017 (In Million Units). 2018. Available online: https://www.statista.com/statistics/263437/global-smartphone-sales-to-end-users-since-2007/ (accessed on 15 December 2018).
10. Newzoo. Top 50 Countries by Smartphone Users and Penetration. 2017. Available online: https://newzoo.com/insights/rankings/top-50-countries-by-smartphone-penetration-and-users/ (accessed on 15 December 2018).
11. Business Insider. The 10 Best Smartphones You Can Buy Right Now—Ranked by Price. Available online: https://www.businessinsider.nl/10-best-smartphones-you-can-get-2017-2017-12/?international=true&r=USAuthor:EdoardoMaggio (accessed on 27 December 2017).
12. Defra. *An Analysis of the Spectrum of Re-Use. A Component of the Remanufacturing Pilot, for Defra, BREW Programme*; Oakdene Hollins Ltd.: Aylesbury, UK, 2007. Available online: http://www.remanufacturing.org.uk/pdf/story/1p374.pdf (accessed on 15 December 2018).
13. Liao, C.H.; Zhang, Y.B. The estimation study on the amount of waste mobile phones in China. *Ecol Econ.* **2012**, *3*, 124–126. (In Chinese)
14. Sarath, P.; Bonda, S.; Mohanty, S.; Nayak, S.K. Mobile phone waste management and recycling: Views and trends. *Waste Manag.* **2015**, *46*, 536–545. [CrossRef] [PubMed]
15. Sabbaghi, M.; Behdad, S. Consumer decisions to repair mobile phones and manufacturer pricing policies: The concept of value leakage. *Resour. Conserv. Recycl.* **2018**, *133*, 101–111. [CrossRef]
16. Vogtlander, J.G.; Scheepens, A.E.; Bocken, N.M.; Peck, D. Combined analyses of costs, market value and eco-costs in circular business models: Eco-efficient value creation in remanufacturing. *J. Remanuf.* **2017**, *7*, 1–7. [CrossRef]
17. Hobson, K.; Lynch, N.; Lilley, D.; Smalley, G. Systems of practice and the Circular Economy: Transforming mobile phone product service systems. *Environ. Innov. Soc. Transit.* **2018**, *26*, 147–157. [CrossRef]
18. Riisgaard, H.; Mosgaard, M.; Zacho, K.O. Local Circles in a Circular Economy-the Case of Smartphone Repair in Denmark. *Eur. J. Sustain. Dev.* **2016**, *5*, 109.
19. Watson, D.; Gylling, A.C.; Tojo, N.; Throne-Holst, H.; Bauer, B.; Milios, L. *Circular Business Models in the Mobile Phone Industry*; Nordic Council of Ministers' TemaNord Report 2017; Nordic Council of Ministers: Copenhagen, Denmark, 2017; p. 560.
20. Wieser, H.; Tröger, N. Exploring the inner loops of the circular economy: Replacement, repair, and reuse of mobile phones in Austria. *J. Clean. Prod.* **2018**, *172*, 3042–3055. [CrossRef]

21. Cao, J.; Chen, Y.; Shi, B.; Lu, B.; Zhang, X.; Ye, X.; Zhai, G.; Zhu, C.; Zhou, G. WEEE recycling in Zhejiang Province, China: Generation, treatment, and public awareness. *J. Clean. Prod.* **2016**, *127*, 311–324. [CrossRef]
22. Greenpeace. Greenpeace Global Mobile Survey 2016. Available online: http://opendata.greenpeace.org/dataset/global-mobile-survey (accessed on 15 December 2018).
23. Deloitte. Global Mobile Consumer Survey—Dutch Edition. 2017. Available online: https://www2.deloitte.com/content/dam/Deloitte/nl/Documents/technology-media-telecommunications/2017%20GMCS%20Dutch%20Edition.pdf (accessed on 15 December 2018).
24. Statcounter. Mobile Vendor Market Share Poland 2017–2018. 2018. Available online: http://gs.statcounter.com/vendor-market-share/mobile/poland (accessed on 15 December 2018).
25. Xu, C.; Zhang, W.; He, W.; Li, G.; Huang, J. The situation of waste mobile phone management in developed countries and development status in China. *Waste Manag.* **2016**, *58*, 341–347. [CrossRef] [PubMed]
26. Porter, M.E. The competitive advantage of nations. *Compet. Intell. Rev.* **1990**, *1*, 14. [CrossRef]
27. Snowdon, B.; Stonehouse, G. Competitiveness in a globalised world: Michael Porter on the microeconomic foundations of the competitiveness of nations, regions, and firms. *J. Int. Bus. Stud.* **2006**, *37*, 163–175. [CrossRef]
28. Rugman, A.M. Porter takes the wrong turn. *Bus. Q.* **1992**, *56*, 59–64.
29. Dunning, J.H. Internationalizing Porter's diamond. *Manag. Int. Rev.* **1993**, *33*, 7–15.
30. Waverman, L. *A Critical Analysis of Porter's Framework on the Competitive Advantage of Nations*; Alan, M.R., Julien, V.D.B., Alain, V., Eds.; Beyond the Diamond (Research in Global Strategic Management, Volume 5); Emerald Group Publishing Limited: Bingley, UK, 1995; pp. 67–95.
31. Gunter, J. Circular Economy isn't just Recycling Products; Repair and Reuse Are also Vital. The Guardian. Available online: www.theguardian.com/sustainable-business/circular-economy-recyling-repair-reuse (accessed on 4 December 2013).
32. Benton, D.; Coats, E.; Hazell, J. *A Circular Economy for Smart Devices: Opportunities in the US, UK and India*; Green Alliance: London, UK, 2015.
33. Kang, H.; Schoenung, J.M. Electronic waste recycling: A review of US infrastructure and technology options. *Resour. Conserv. Recycl.* **2005**, *45*, 368–400. [CrossRef]
34. Kahhat, R.; Kim, J.; Xu, M.; Allenby, B.; Williams, E.; Zhang, P. Exploring e-waste management systems in the United States. Resources. *Conserv. Recycl.* **2008**, *52*, 955–964. [CrossRef]
35. Huang, K.; Guo, J.; Xu, Z. Recycling of waste printed circuit boards: A review of current technologies and treatment status in China. *J. Hazard. Mater.* **2009**, *164*, 399–408. [CrossRef]
36. Ghisellini, P.; Cialani, C.; Ulgiati, S. A review on circular economy: The expected transition to a balanced interplay of environmental and economic systems. *J. Clean. Prod.* **2015**, *114*, 11–32. [CrossRef]
37. King, A.M.; Burgess, S.C.; Ijomah, W.; McMahon, C.A. Reducing waste: Repair, recondition, remanufacture or recycle? *Sust. Dev.* **2006**, *14*, 257–267. [CrossRef]
38. Stahel, W.R. *The Utilization Focused Service Economy: Resource Efficiency and Product Life Extension, in The Greening of Industrial Ecosystems*; National Academy Press: Washington, DC, USA, 1994; pp. 178–190.
39. Ijomah, W. A Model—Based Definition of the Generic Remanufacturing Business Process. Ph.D. Thesis, University of Plymouth, Plymouth, UK, 2002.
40. Ijomah, W.L.; Chiodo, J.D. Application of active disassembly to extend profitable remanufacturing in small electrical and electronic products. *Int. J. Sustain. Eng.* **2010**, *3*, 246–257. [CrossRef]
41. Guest, G.; Bunce, A.; Johnson, L. How Many Interviews Are Enough? An Experiment with Data Saturation Variability. *Field Methods* **2006**, *18*, 59–82. [CrossRef]
42. Galvin, R. How many interviews are enough? Do qualitative interviews in building energy consumption research produce reliable knowledge? *J. Build. Eng.* **2015**, *1*, 2–12. [CrossRef]
43. Brix-Asala, C.; Geisbüsch, A.K.; Sauer, P.C.; Schöpflin, P.; Zehendner, A. Sustainability Tensions in Supply Chains: A Case Study of Paradoxes and Their Management. *Sustainability* **2018**, *10*, 424. [CrossRef]
44. Charter, M. *Designing for the Circular Economy*; Charter, M., Ed.; Routledge: Abingdon, NY, USA, 2018.
45. Charter, M. Repair Cafes. *J. Peer Prod.* **2018**, *3*. Available online: http://peerproduction.net/wp-content/uploads/2018/07/jopp_issue12_charter.pdf (accessed on 15 December 2018).

Article

The Role of Cities in the Sharing Economy: Exploring Modes of Governance in Urban Sharing Practices

Jenny Palm, Karolina Södergren * and Nancy Bocken

International Institute for Industrial Environmental Economics (IIIEE), Lund University, P. O. Box 196, 221 00 Lund, Sweden; jenny.palm@iiiee.lu.se (J.P.); nancy.bocken@iiiee.lu.se (N.B.)

* Correspondence: karolina.sodergren@iiiee.lu.se; Tel.: +46-46-222-00-00

Received: 12 November 2019; Accepted: 10 December 2019; Published: 12 December 2019

Abstract: Cities have for a long time been key actors in sustainable urban development, and in recent times, also for the sharing economy, as they provide a fertile breeding ground for various sharing initiatives. While some of these initiatives build on existing practices and infrastructures such as public libraries and repair workshops, others require the involvement of private companies, as in the case of car sharing. The sharing economy might therefore require a significant reinterpretation of the role of local governments, businesses and citizens, which in turn might imply a complex re-organisation of governing. This article will explore what potential roles cities might have in governing the sharing economy. Four Swedish cities serve as case studies for this purpose: Stockholm, Gothenburg, Malmö and Umeå. City data was collected primarily through qualitative means of investigation, including workshops, interviews and desk research. In Malmö, additional participatory observations were conducted on the testbed Sege Park. Results were analysed with a framework developed for understanding the various governing roles for cities in the sharing economy. Three dominant modes of governing were identified and discussed: governing by provision and authority; governing by partnership and enabling; and governing through volunteering. The four cities made use of all three governing modes, although with a primary focus on governing by authority and governing through partnership. When characterised by governing through volunteering, projects were always initiated by the city, but then run formally by an NGO. While all governing modes may have a role and a purpose in the sharing economy, it is still important that cities reflect upon what are their actual implications. Risks include a collaborative governing mode out-competing some businesses, for example, and a self-governing mode reducing the action space of the volunteer sector.

Keywords: sharing economy; sharing cities; sustainable urban governance; sharing business models; sustainable business models

---

1. Introduction

Cities have long been key actors as planners and governors for sustainable urban development [1]. More recently, cities have also become a breeding ground for new sharing economy initiatives [2]. In a sharing economy, under-utilised assets are shared in order to optimise resource use. Individuals and families rent or borrow their assets to other people, which can be contrasted with a gig economy in which individuals offer their services to companies on a part-time basis [3].

Sharing platforms are tools that enable citizens to share, lend, sell and rent resources. Such sharing activities have the potential to promote more efficient use of resources and reduce the environmental impacts from consumption [4,5]. The new frontier in the sharing economy is called 'prosumers', and commonly refers to the formation of energy sharing communities between local inhabitants and utility companies [6,7]. The potential for sustainability of the sharing economy is still debated and needs further investigation [8]. This, however, will not be the focus for this paper. Rather, we zoom in on the different roles that cities may adopt in governing the sharing economy.

In some cases, cities may be overwhelmed by the emergence of a great variety of start-ups, e.g., in the area of mobility (e.g., car sharing, taxi services), and are forced to catch up by developing policies encouraging or inhibiting the emergence of such services. Other cities take a much more proactive role by shaping the new sharing initiatives. Examples of the latter involve the establishment of tool libraries or tool pools where people may borrow tools at no or low cost [9,10]. In this article, we will analyse the different roles local governments have taken in the sharing economy in four Swedish cities.

The potential for cities to leverage different resource-efficient sharing initiatives is potentially high, but at the same time not limitless. Local governments have authority in a restricted geographic area, with limited space for infrastructure, initiatives and services. Local governments need to work in governance processes, made up of partnerships and network-orientated decision-making in a sophisticated interplay between public, private and non-profit organisations [11]. Cities need to mobilise external actors (and their resources) for the formulation and implementation of complex cross-boundary issues such as sustainability, circular economy, sharing and experimentation [12]. Cities are encouraged to be at the forefront of these processes, and when doing so, the interests of external actors need to be integrated into local policy-making. At the same time, cities still need to carry out their compulsory duties and fulfil their responsibilities for providing welfare services to their citizens. Engagement in experimentation and sharing, for example, then takes resources from the cities and influences how they can perform their regulatory tasks. Acquier et al. [13] discuss the need to understand the complex and sometimes contradictory objectives cities have when advising them on how to engage in the sharing economy. Cities are supposed to navigate between requirements such as acting in the interest of the public and working on digital platforms without causing any negative externalities [13].

On the one hand, it is not obvious that a local government should engage in the sharing economy. It might be better to leave this space open for private and voluntary initiatives to enter. On the other hand, in countries like Sweden, where the voluntary sector has traditionally focused on sports and leisure, this means the sharing economy might never take off [6]. Yet, the sharing economy could have many benefits for municipal development and democratisation [14]. May et al. [15], for example, argue that digital sharing platforms could be a way for cities to pursue a more attractive image. This, however, requires a reinterpretation of the role of local governments, businesses and citizens, which in turn implies a complex re-organisation of governing. Cities might need to explore new modes of governance [16].

In this article, we address the following research question: What potential roles do cities have in governing the sharing economy? The dilemmas faced by cities when engaging in the sharing economy are described, and an analytical framework for discussing what sharing issues municipalities should engage in, when and why, is presented.

2. Background

This section discusses the sharing economy in a local context, followed by a background on how to study the governing of the sharing economy.

2.1. Sharing Economy: Background and Local Contexts

While not a novel phenomenon per se [17], sharing has gained popularity in recent years through the ICT and social networks (e.g., digital platforms), enabling a sharing economy [15,18]. This has been called an "idea that will change the world" through its potential contribution to future economic activity [19].

Within the sustainability discourse, the sharing economy has been endorsed as a means for dematerialising our economies [2,20]. In the background lie growing concerns about the unsustainable consumption patterns characterising our times and the resulting over-exploitation of natural resources that lead to increasing pollution, climate change, biodiversity loss, etc. [2,21].

In academia, the sharing economy has mainly been studied as a solution that leverages the idling capacity of goods and services by offering access over ownership [2]. Critics, however, claim that it is becoming a 'catch-all term', and call for a more detailed definition [22], notably because semantic confusion can lead to the concept being misused in ways that go against its intended meaning. Similar to the term 'green washing', 'share washing' has been used to describe such situations in which companies make illicit linkages between their core business and sharing as a sustainable consumption practice [2]. 'Platform capitalism' or 'neoliberalism on steroids' are other phrases that have been used to describe the controversy surrounding the definition of the sharing economy [23].

Taking the above into account and contrasting the sharing economy with terms like collaborative consumption, the gig economy, peer-to-peer economy, circular economy, and consumer-to-consumer economy, Acquier et al. [24] proposed an organising framework for the sharing economy that rests upon "three foundational cores". Visible in Figure 1 below, these are: (1) access economy (e.g., the sharing of underutilised assets for optimising resource use), (2) platform economy (e.g., digital platforms mediating decentralised exchanges among peers), and (3) community-based economy (e.g., coordination through "non-contractual, non-hierarchical or non-monetized forms of interaction") [24].

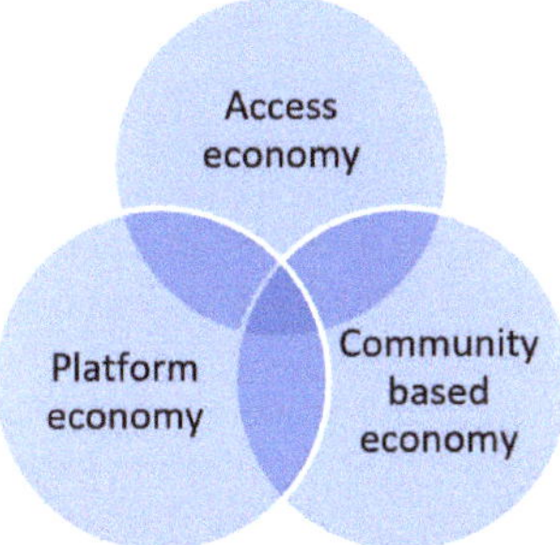

Figure 1. Conceptual underpinnings of the sharing economy. Borrowed from Acquier et al. [24].

Cities make up a particularly fertile breeding ground for sharing economy initiatives [25]. Davidson and Infranca [26] discussed how proximity and density features of urban environments bring about, e.g., agglomeration benefits. Reduced transportation costs and access to a diverse labour market are some of the factors that allow for increased productivity and economic growth in cities. A deeper pool of sellers and buyers foster specialisation, and thus the provision of goods and services that may otherwise not exist locally. The development of new products and services is also linked to the rapid information exchanges that are enabled by modern infrastructure in cities, which in turn help spur innovation.

Providing a similar logic, May et al. [15] stated that the "increasing urbanization and densification of the population helps smooth the friction of the sharing economy". In addition, conversely, according to Davidson and Infranca [26] the sharing economy is also "an agent of urban transformation". This is due to its potential impact on the economic, physical, geographical and social landscapes in cities. Hofman et al. [27] believe that the public sector has mainly acted as a regulatory body in the sharing economy, neglecting other potential roles as customers and platform providers. They undertake a theoretical analysis of how the different roles a city can take in the sharing economy relate to four different values: professionalism, efficiency, service, and engagement. They find opportunities and challenges of each role for the four public values, but conclude that there is a need for empirically based research on the subject. This article seeks to contribute in this exact regard.

Across the globe, cities have taken different approaches to the sharing economy, with regard to both its opportunities and challenges. Developments are emerging from collaborative initiatives such as the 'sharing cities' network, culminating in November 2018 with a Sharing Cities Declaration [10]. In Sweden, the sharing economy has been described as fairly under-developed as compared to other countries [15]. The majority of companies working in the sharing economy are still in a start-up phase, and thus have a low level of revenue. That said, Sweden is seen to have good conditions for furthering and benefitting from the sharing economy. Key factors include high levels of employment, innovative entrepreneurship, specialisation in IT and communication services, use of internet and mobile phones, and strong public interest in sustainability related matters [18]. This is reflected in the mapping and analysis of sharing services and initiatives in Swedish cities provided by Markendahl et al. [10], as well as in the constant increase of digital sharing platforms operating within national borders. A 2016 survey undertaken by Nordea also showed an annual increase in sharing economy 'users' of 3%, reaching approximately 13% of the Swedish population [18].

2.2. Governing the Sharing Economy

Palm et al. [9] claimed that achieving sustainable consumption required a reinterpretation of the role of public organisations, businesses and citizens because of the "complex challenges and institutional contradictions for governance" that it entails. This is in line with previously developed theories on the multi-level governance and dispersion of power that characterises our modern society [28–30].

In relation to governing the sharing economy, the need for "tougher rules" has been expressed in, e.g., The Guardian [31]. Lacking are policy guidelines ensuring that both providers and consumers are protected in sharing activities. Indeed, Cannon and Summers [32] went as far as arguing that "regulation is often the most significant barrier to future growth for sharing economy firms".

Murillo et al. [23] believe that the sharing economy should be "of particular interest to governments and public authorities since it is estimated that 70% of Europeans and 72% of Americans are involved in SE [sharing economy] activities" (p. 66). Highlighting situations in which national governments fail to establish an appropriate regulatory environment, however, they notice that "local governments are the ones that must carry the burden of taxing firms and enforcing law" (p. 70). This is also recognised by Davidson and Infranca [26] when discussing the sharing economy as an urban governance challenge, pointing to the intrinsic dependences between sharing activities and specific urban conditions.

Indeed, local or "city governments around the world are increasingly adopting policies to regulate some forms of sharing" [14]. Building upon a framework for urban climate governance developed by Bulkeley and Kern [12], Zvolska et al. [14] suggested the different roles that cities assume when governing urban sharing are: governing by regulation, governing by provision, governing by enabling, and governing by consumption. We will come back to this next, when discussing our analytical framework.

3. City Engagement in the Sharing Economy—An Analytical Framework

To be able to discuss city engagement in the sharing economy, we develop an analytical framework that can help with this examination. The framework is inspired by Bulkeley and Kern [12] and Zvolska et al. [14]. The extended typology includes five different modes of governing based on the type of capacity cities can have or take in different settings:

- Self-governing or governing by example, which relies on the organisational capacity of the municipality to manage its own operations;
- Governing by provision, which is related to the municipal role as provider of different goods and services;
- Governing by authority, which concerns municipal ability to mandate a certain behaviour and impose sanctions if such a mandate is not followed.

- Governing by enabling, which refers to the municipality's capacity to persuade and encourage behaviour through the use of positive incentives such as subsidies, information campaigns, and facilitation of different kinds of initiatives.
- Governing by partnership, which is characterised by an equal relationship between the municipality and other actors. Cities are one among several actors in a project/partnership, and has no formal steering power over the other members.

In Sweden, a different model has been used in the project "Malmö Innovation Arena" to discuss what role the city could and should adopt in relation to innovations. In Malmö Innovation Arena, focus has been placed on when and how much the city of Malmö should engage in innovation processes, and when to stay out and give space for other actors to take the lead. The important parameters in this model are related to three different groups of actors (the city, businesses and volunteer organisations) as well as to who initiates, decides and implements various solutions. Table 1 presents a combination of the frameworks and ideas suggested by Bulkeley et al., Zvolska et al., and Malmö Innovation Arena.

Table 1. Governing roles for cities in the sharing economy.

	Governing by Provision and Authority	Governing by Partnership and Enabling	Governing through Volunteering
Initiator	Local government/authorities	Local government together with others	Citizens or Non-governmental organisations (NGOs)
Decision made by	Local government	Shared between involved partners	By the initiator
Activities run by	Local government	Local government and partner	By the initiator
The public sector has	Steering role	Cooperative role	Role when needed to stimulate and facilitate implementation

In a process characterised by governing by provision and authority, the city initiates the issue at hand. The results can be defined from the beginning of a project, resources are calculated, and all decisions are made within the city administration.

In governing by partnership and enabling, the issue is initiated and run collaboratively with others. From the city's point of view, the outcome is more open-ended, decision-making is done in networks of public-private partnership and implementation in collaboration with all stakeholders.

In governing through volunteers, the results of a project are uncertain, and not easily predictable by the city administration. The process lies in the hands of other actors and the role of the city is mainly to support or facilitate these processes. Reasons for cities to engage in volunteer initiatives include that the cities' financial or social vitality is strengthened, the accessibility to public spaces is increased, as well as the quality in the location in which the initiative is implemented. The initiative, however, is never launched by the city, and always by private actors and/or civil society.

Below, this framework will be used to analyse how four Swedish cities have considered their participation in the sharing economy.

4. Methods

The data collection was conducted as part of Sharing Cities Sweden (www.sharingcities.se), which is an initiative that aims to "develop world-leading test-beds for the sharing economy in Stockholm, Gothenburg, Malmö and Umeå". Qualitative methods were used to gather research data, including workshops, semi-structured interviews and participant observation.

For each city, we started by collecting documents, articles and policy papers related to sharing. We then conducted a workshop with representatives from all four cities in April 2019, and followed up with more detailed interviews in September and November 2019.

The workshop was organised as part of the Lund Sustainability Week on 8 April 2019. The whole event gathered over 80 participants from four Swedish cities—Malmö, Gothenburg, Stockholm and Umeå—as well as academics and other stakeholders interested in Sharing Cities. In the morning, different sharing projects were presented. In the afternoon, twenty participants were invited from Umeå, Gothenburg, Stockholm, and Malmö cities for a closed workshop held partly in English (moderation) and partly in Swedish (subgroup discussion). During the workshop, participants were grouped city-wise, i.e., representatives from the same city were placed in the same group. Each group was chaired by a researcher who also took notes from the discussions. Four main questions were discussed:

1. the role of the city in the sharing economy
2. the city's ambitions in the sharing economy
3. business models and future opportunities
4. the role of digitalisation in the sharing economy

We also conducted complementary interviews with representatives from the three cities, to follow up on how they work with sharing initiatives (Gothenburg, Malmö and Umeå). The interviews were semi-structured covering the following themes: Sharing projects, content; Engaged actors/organisations; What does the engagement look like, mechanisms for engagement (facilitating and conducting experimentation, taxes, regulations, procurement, partnerships, financing, etc.); Problems; Possibilities; What do you envision the city's role in the future; and Other.

Additionally, in Malmö, participatory observations have been done at 24 dialogue meetings in their testbed Sege Park since June 2018. According to the city plans, there will be up to 900 dwellings in Sege Park by 2025. The development process is supposed to be done using a sustainable approach with a specific focus on creating a low-carbon district and innovative sharing solutions. The researchers have participated in 24 dialogue meetings where property developers and city representatives met to discuss how to develop the area and what solutions to invest in. During the meetings, field notes were written, addressing the dialogues between participants. Fifteen interviews were also conducted focusing on the collaborative process in Sege Park.

The interview questions in Sege Park were semi-structured and covered the following themes: the interviewees' professional background and current professional role, the collaborative planning process and how it was (or was not) related to the work processes in Sege Park, experiences so far with the process and descriptions of how the process and included issues discussed at the meeting played out, interaction between the participants and finally, which parts had worked well and what improvements could potentially be made. A summary of the methodological approach used for this research is available in Figure 2 below.

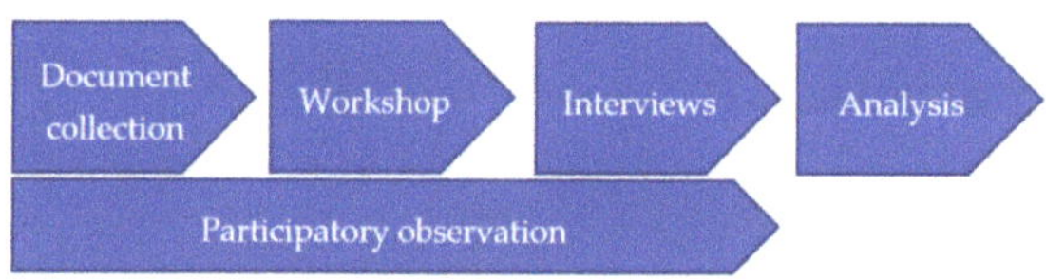

Figure 2. Methods applied for this research.

We will now use examples of city engagements to discuss how the cities are governing the sharing economy, as well as how they think about it.

5. Results: Different City Roles in the Sharing Economy

Swedish cities have invested in various solutions to enable their citizens to move towards sharing. Examples of how cities contribute through different governing modes have been discussed in an earlier contribution [9]. In this study, we will analyse how cities regard their roles in the sharing economy. The results are organised according to the three governing modes presented in the analytical framework above.

5.1. Governing by Self-Government, Provision and Authority

Governing by self-government, provision and authority includes activities where the city manages its own operations, provide goods or services or uses sanctions, for example, to mandate a behaviour.

During the workshop, all four cities mentioned that it was important that they adopt a role as a regulator to be able to attain a quick transformation towards the sharing economy. By taking an authoritative role, the city can "shape" the urban environment, as one of the representatives expressed it. The cities used public procurement actively for this purpose. For example, car sharing was procured by the cities as a way to contribute to a sharing economy as well as to environmental sustainability with the assumption that it reduces the numbers of cars needed by the city.

During the workshop, city planning was emphasised as an important way for all the cities to contribute to a sustainable and circular society. Not only car-free districts where mentioned, but also the possibility of planning for mobility as a service (MAAS) by including transportation hubs in the city planning process. Umeå thought that it was important for the city to also be engaged through its authoritative function to be able to guarantee that sharing was something that *all* citizens could take part of. Umeå had found it easier to establish private sharing solutions such as carpooling in middle-class areas, compared to areas with socio-economic problems. Private carpooling companies hesitated to park their cars in less wealthy areas, because of bad experience with demolished cars. This experience was shared by Stockholm and Malmö. An important role for the city, which was highlighted during the workshop, was to make sure that all citizens benefitted from sharing services. Otherwise, the risk is that only citizens in favourable socio-economic situations would be able to take part in the sharing economy. From this perspective, the city representatives thought that the city should use authoritative means such as planning and public procurement to contribute to the development of a local sharing economy.

As mentioned above, Sege Park in Malmö is a testbed for sharing solutions. Over 30 sharing solutions have been up for discussion so far (field notes 28 February 2018). When it comes to the governing process related to these sharing solutions, it is not so clear what governing approach the city of Malmö has used. The process has been characterised by all three approaches (self-government, collaborative and volunteers). We will address the collaborative and volunteer aspects further down. In relation to governing by self-government, provision and authority, the city of Malmö has regulated that sharing should be a main characteristic of Sege Park. This was an explicit requirement in the land allocation process that property developers needed to take into consideration. In the city's sustainability strategy, it says that "Sege Park shall be a testbed for how sharing solutions can be used in the city district area". In the land allocation process, the property developers needed to show how they planned for sharing services, such as having premises that could be used for sharing in buildings, car and electric bike pools, outdoor barbecue "kitchens" to be shared, etc. The city used the zoning plan to create a car-free neighbourhood, which also encouraged public transportation and car sharing.

However, the city has not always been prepared to support sharing solutions even if they had the authority to do so. Sege Park in Malmö has a sustainability approach, and the property developers in the area planned for installing PVs on the roof and on the ground, and for developing a micro grid which would make it possible to share electricity among the different property developers. The city of Malmö was not prepared to distribute space in the zooning plan for PVs in the green areas that the city owned and controlled. Moreover, in relation to the micro grid, the city chose not to act. For regulative reasons, it is not possible to have a micro grid where you transport electricity between different buildings if the buildings have different owners, as is the case in Sege Park. The city of Malmö could, however, have asked the Swedish Government if Sege Park could be a testbed for trying this out. However, they chose not to.

To conclude, the cities have applied governing by self-government, provision and authority to enhance a sharing economy. There is, however, also one example of when they deliberately chose not to use this governing mode, and thereby, hindered the implementation of different sharing solutions.

5.2. Governing by Partnership and Enabling

Governing by partnership and enabling is characterised by cooperation between the municipality and other actors, as well as by the city using positive incentives to encourage a behaviour.

This was the mode that the four cities highlighted as the most prominent one for developing a sharing economy. In general, the city was seen to have a key role in facilitating and enabling various sharing initiatives. They emphasised their role as owners of certain sharing initiatives such as "smartakartan.se" (the smart map), which is a digital map highlighting what kind of sharing solutions are available in the city and where they are located.

All four cities also agreed that digitalisation was a critical enabling factor for sharing, through apps, digital platforms, etc. The cities felt that they had an important role in developing those ICT tools, but also that it required many man-hours for development and maintenance. This time-consuming work was seen as difficult for a private actor to engage in, which more or less forced the cities to take a role in the development process. Swedish cities also have a lot of data and statistics stored, such as travel patterns and socio-economic data, which is relevant for many sharing services. The cities were willing to share this type of data, as long as it did not violate any regulations.

In Sege Park, Malmö city emphasised the collaborative governing mode from the start of the project and made clear that in order to fulfil the goal of Sege Park as a testbed for sharing, all actors involved needed to contribute.

The city was prepared to facilitate sharing solutions through planning, organising workshop and study visits, inviting experts and practitioners, and providing support with finding funding to try out different sharing solutions. The building developers needed to decide what sharing activities they wanted to invest in and how they would collaborate among themselves around different sharing services. They had to fill out a questionnaire mentioning what kind of sharing services they had in mind, how these would be organised and paid for, and if the services would be available for their tenants only or for anyone.

This collaborative mode worked well for some property owners, while others chose not to participate. The reason for not contributing was that the developers believed that they had been allocated small land areas, which according to them made it impossible to be involved in the sharing solutions discussed:

> *We do not have the possibility to work with sharing spaces because we have not so much land allocated to us. Thus, so far we have not been able to contribute and we will continue to have less opportunities to participate in sharing solutions. (Property developer 7)*

As seen in the quotation above, although not the case, this property developer had interpreted discussions in a way that suggested that all sharing solutions need space.

Other developers felt that too much time was spent on developing sharing solutions, "without guidance or direction". They lacked concrete examples on successful sharing solutions and thought that sharing services came with risks that were too great.

> *Sharing has not been discussed in relation to existing sharing solutions in other areas. /… / We have a list with sharing solutions, but we have not discussed how this will work in practice. (Property developer 9)*

The property developers lacked evaluations and wanted more knowledge about what the results have been in other areas where sharing solutions have already been implemented.

Several of the developers also questioned the idea that building developers and the city should plan for future sharing services for residents without having any knowledge of future demand:

> *This is the difficult part with sharing. We cannot force the future residents to use them. And it is hard to know what the future demand for sharing solutions will consist of. We can prepare for sharing, but not decide what the future residents will do. (Property developer 2)*

Another property developer said:

> *The sharing group becomes speculative. We spend time on things where we don't know if this is the right solution to work with. We don't know which sharing solution will be the one asked for. (Property developer 12)*

These property developers thought that sharing services should be developed by the citizens in relation to their demand, which also takes us to the next governing mode.

5.3. Governing through Volunteering

Governing through volunteering emphasises the role of NGOs and citizens, and that it is volunteers who should take a lead in developing the sharing economy.

During the workshop with the cities, a perspective that was often put forward was the importance of the sharing economy as led by the demands from the users. As described above, it was, however, a critique in Sege Park that the citizens were not involved in the development of the sharing economy. The potential result, highlighted by actors in Sege Park, could be that sharing solutions were developed without any existing demand. Another problem could be that commercial interests conquer the area, and leave no room for other actors. This was discussed in a debate article recently published by one of Sweden's biggest newspaper, Dagens Nyheter [33]. In the article, several volunteer organisations came together and criticised Swedish municipalities for out-competing NGO activities, including environmental, social and human rights work to promote social or political change. They felt that when Swedish municipalities start to run repair workshops together with café services, they outcompete "the social and local economy", referring to the volunteer sector and NGOs.

A challenge raised multiple times by the cities during the workshop was related to finding sustained financial resources for initiatives with volunteer engagement. Sharing experiments that were done with volunteers were considered short-time engagements. At the same time, there were several examples of successful sharing services where the civil society had a leading role, for example a toy library in Gothenburg and clothing exchange days in Malmö. These projects favour social inclusion as well as sustainability, according to the cities.

Gothenburg emphasised that they wanted the "local society" to provide social services. They saw the role of the city as mainly supportive, and Gothenburg was the city that reflected most upon governing through volunteer mode during the workshop. They believed that the role of the city was to provide a kind of 'payback' for social services, which would give more security to private initiatives.

6. Conclusions

The aim with this article was to increase the understanding of the different roles a city might take in governing the sharing economy. The idea was not to evaluate or assess how the cities have performed, but to identify and reflect upon different governing modes used by local governments.

The four Swedish cities that were analysed relate to all three governing modes that were defined: governing by provision and authority; governing by partnership and enabling; and governing through volunteering. However, in reality, examples characterised by authority and partnership modes of governance dominated the discussions. One possible explanation for this is that a high involvement of the cities is needed initially, to transform the economy to a sharing economy. A strong actor, such as a city, might be needed to lead this process, in particular when private companies compete for space and market share. Indeed, the proliferation of scooters and bike sharing schemes within city boundaries pursued by private companies indicates that intervention and more of an authority and partnership role adopted by the cities would be desirable. May et al. [15] highlight 'trust' as a critical lever for making the sharing economy work. Swedish cities as public actors are usually seen as highly trustworthy and have been important trust creators for many experimental projects, such as in the sharing test beds in the four cities [9].

Having strong cities leading the transformation towards a sharing economy also comes with certain problems. When cities take the lead in the sharing economy, there is a risk that they out-compete the volunteer sector. It was expressed during interviews and in the workshop that when governing through volunteering, cities usually initiate the sharing solution and then hand over to an NGO or an individual to take care of the day-to-day activities. The cities' experience it to be difficult to sustain sharing activities where volunteers play a major role. An explanation for this lack of perseverance could be that the cities initiated solutions that did not attract enough interest from the public, contributing to the engagement quickly fading out. The role of volunteers and how it relates to the role of cities needs further investigation in the future to understand why volunteers have difficulty in maintaining engagement.

Another problem raised by the property developers in the testbed Sege Park in Malmö was that there may be a mismatch between the sharing services cities offer and the services citizens want. The risk is that the city ends up with many sharing solutions that no one is using. An increased involvement of volunteers could here be a way to close such a potential gap. The strength of the volunteer sector lies in its proximity to citizens as well as in their localised understanding of existing needs. This knowledge might disappear if cities start to dominate the process.

Cities can also resist sharing solutions, like Malmö did when opposing shared PV parks in Sege Park. Cities as barriers for the development of a sharing economy is another perspective that requires further research.

In conclusion, all governing modes are needed in the development of a sustainable sharing economy. For cities, however, it is important to reflect upon what the implications are of different governing modes.

Author Contributions: Responsible for the Introduction were J.P., K.S. and N.B.; Background, K.S.; Analytical framework, J.P.; methodology, J.P., K.S. and N.B.; analysis, J.P. with input from K.S. and N.B.; Conclusions, J.P., K.S. and N.B.; revision, J.P. and K.S.

Funding: This research was funded by VINNOVA, for the Strategic Project Sharing Business Models within the Sharing Cities program, grant number 2018-04646 and by the Swedish Energy Agency project no 46016-1 "Smart symbiosis—collaboration for common resource flows".

Acknowledgments: We would like to acknowledge the support of the Lund University IIIEE teams in facilitating the cities workshop.

Conflicts of Interest: The authors declare no conflict of interest. The funders had no role in the design of the study; in the collection, analyses, or interpretation of data; in the writing of the manuscript, or in the decision to publish the results.

References

1. Fenton, P.; Gustafsson, S.; Ivner, J.; Palm, J. Sustainable energy and climate strategies: Lessons from planning processes in five municipalities. *J. Clean. Prod.* **2015**, *98*, 213–221. [CrossRef]
2. Curtis, S.K.; Lehner, M. Defining the sharing economy for sustainability. *Sustainability* **2019**, *11*, 567. [CrossRef]
3. De Stefano, V. The rise of the just-in-time workforce: On-demand work, crowdwork, and labor protection in the gig-economy. *Comp. Labor Law Policy J.* **2015**, *37*, 471. [CrossRef]
4. Martin, C.J.; Upham, P.; Klapper, R. Democratising platform governance in the sharing economy: An analytical framework and initial empirical insights. *J. Clean. Prod.* **2017**, *166*, 1395–1406. [CrossRef]
5. Frenken, K. Sustainability perspectives on the sharing economy. *Environ. Innov. Soc. Transit.* **2017**, *23*, 1–2. [CrossRef]

6. Magnusson, D.; Palm, J. Come together-the development of Swedish energy communities. *Sustainability* **2019**, *11*, 1056. [CrossRef]
7. Kooij, H.-J.; Oteman, M.; Veenman, S.; Sperling, K.; Magnusson, D.; Palm, J.; Hvelplund, F. Between grassroots and treetops: Community power and institutional dependence in the renewable energy sector in Denmark, Sweden and the Netherlands. *Energy Res. Soc. Sci.* **2018**, *37*, 52–64. [CrossRef]
8. Schor, J. Debating the sharing economy. *J. Self-Gov. Manag. Econ.* **2016**, *4*, 7–22.
9. Palm, J.; Smedby, N.; McCormick, K. The role of local governments in governing sustainable consumption and sharing cities. In *A Research Agenda for Sustainble Consumption Governance*; Mont, O., Ed.; Edward Elgar Publishing: Cheltenham, UK, 2019.
10. Markendahl, J.; Hossain, M.I.; Mccormick, K.; Lund, T.; Möller, J.; Näslund, P. Analysis of sharing economy services: Initial findings from sharing cities Sweden. *Nord. Balt. J. Inf. Commun. Technol.* **2018**, *2018*, 239–260.
11. Wihlborg, E.; Palm, J. Who is governing what? Governing local technical systems—An issue of accountability. *Local Gov. Stud.* **2008**, *34*, 349–362. [CrossRef]
12. Bulkeley, H.; Kern, K. Local government and the governing of climate change in Germany and the UK. *Urban Stud.* **2006**, *43*, 2237–2259. [CrossRef]
13. Acquier, A.; Carbone, V.; Massé, D. How to create value (s) in the sharing economy: Business models, scalability, and sustainability. *Technol. Innov. Manag. Rev.* **2019**, *9*, 5–24. [CrossRef]
14. Zvolska, L.; Lehner, M.; Voytenko Palgan, Y.; Mont, O.; Plepys, A. Urban sharing in smart cities: The cases of Berlin and London. *Local Environ.* **2019**, *24*, 628–645. [CrossRef]
15. May, S.; Konigsson, M.; Holmstrom, J. Unlocking the sharing economy: Investigating the barriers for the sharing economy in a city context. *First Monday* **2017**, *22*. [CrossRef]
16. Evans, B.; Joas, M.; Sundback, S.; Theobald, K. Governing local sustainability. *J. Environ. Plan. Manag.* **2006**, *49*, 849–867. [CrossRef]
17. Belk, R. Sharing. *J. Consum. Res.* **2009**, *36*, 715–734. [CrossRef]
18. SOU. Delningsekonomi på användarnas villkor. In *Swedish Governments Official Reports*; 2017:26; Wolters Kluwers: Stockholm, Sweden, 2017.
19. Gregory, A.; Halff, G. Understanding public relations in the 'sharing economy'. *Public Relat. Rev.* **2017**, *43*, 4–13. [CrossRef]
20. Boons, F.; Bocken, N. Towards a sharing economy–Innovating ecologies of business models. *Technol. Forecast. Soc. Chang.* **2018**, *137*, 40–52. [CrossRef]
21. Heinrichs, H. Sharing economy: A potential new pathway to sustainability. *Gaia-Ecol. Perspect. Sci. Soc.* **2013**, *22*, 228–231. [CrossRef]
22. Belk, R. Sharing versus pseudo-sharing in Web 2.0. *Anthropologist* **2014**, *18*, 7–23. [CrossRef]
23. Murillo, D.; Buckland, H.; Val, E. When the sharing economy becomes neoliberalism on steroids: Unravelling the controversies. *Technol. Forecast. Soc. Chang.* **2017**, *125*, 66–76. [CrossRef]
24. Acquier, A.; Daudigeos, T.; Pinkse, J. Promises and paradoxes of the sharing economy: An organizing framework. *Technol. Forecast. Soc. Chang.* **2017**, *125*, 1–10. [CrossRef]
25. McLaren, D.; Agyeman, J. *Sharing Cities: A Case for Truly Smart and Sustainable Cities*; MIT Press: Cambridge, MA, USA, 2015.
26. Davidson, N.M.; Infranca, J.J. The sharing economy as an urban phenomenon. *Yale Law Policy Rev.* **2015**, *34*, 215.
27. Hofmann, S.; Sæbø, Ø.; Braccini, A.M.; Za, S. The public sector's roles in the sharing economy and the implications for public values. *Gov. Inf. Q.* **2019**, *36*, 101399. [CrossRef]
28. Rhodes, R.A.W. *Understanding Governance: Policy Networks, Governance, Reflexivity and Accountability*; Open University Press: Buckingham, UK, 1997.
29. Pierre, J. *Debating Governance: Authority, Steering, and Democracy*; Oxford University Press: Oxford, UK, 2000.
30. Peters, G.; Pierre, J. Multi-level governance and democracy: A Faustian bargain? In *Multi-Level Governance*; Bache, I., Flinders, M., Eds.; Oxford University Press: New York, NY, USA, 2004; pp. 75–91.
31. Jericho, G. The dark side of Uber: Why the sharing economy needs tougher rules. *The Guardian*, 18 April 2016.

32. Cannon, S.; Summers, L.H. How Uber and the sharing economy can win over regulators. *Harv. Bus. Rev.* **2014**, *13*, 24–28.
33. Stuart Hamilton, U.; Skogar, P. DN debatt—Kommunal dumpning slår sönder den sociala ekonomin. *Dagens Nyheter*, 13 August 2019.

Article

Factorial Decomposition of the Energy Footprint of the Shaoxing Textile Industry

Xiaopeng Wang [1], Xiang Chen [1], Yiman Cheng [1], Luyao Zhou [1], Yi Li [2,3,*] and Yongliang Yang [1,*]

1 Ecological Civilization Research Center of Zhejiang Province, School of Economics and Management, Zhejiang Sci-Tech University, Hangzhou 310018, China
2 East China Sea Institute/Collaborative Innovation Center of Port Economy, Ningbo University, Ningbo 315211, China
3 Fashion Department of International United Faculty between Ningbo University and University of Angers/Faculty of Tourism and Culture, Ningbo University, Ningbo 315201, China
* Correspondence: liyi1@nbu.edu.cn (Y.L.); royyang@zju.edu.cn (Y.Y.); Tel.: +86-574-8760-3821 (Y.L.); +86-571-8684-3676 (Y.Y.)

Received: 8 March 2020; Accepted: 30 March 2020; Published: 3 April 2020

Abstract: To present great environmental pressure from energy consumption during textile production, this paper calculates the energy footprint (EFP) of Shaoxing's textile industry, from 2005 to 2018. Moreover, this study analyzes the relationship between Shaoxing's textile industry energy consumption and economic development by using decoupling theory. Furthermore, the Logarithmic Mean Divisia Index decomposition method was employed to investigate the main factors that affect the EFP of Shaoxing's textile industry. Research results show the following: (1) The growth rate of the total output value of Shaoxing's textile industry was greater than the growth rate of the EFP, from 2005 to 2007. Thus, the decoupling state showed a weak decoupling, and EFP intensity decreased. (2) The EFP and economic growth were mainly based on the strong decoupling of Shaoxing's textile industry from 2008 to 2015 (except for 2011), and EFP intensity declined further. (3) Economic recession in the textile industry was severe in Shaoxing, from 2016 to 2018, and the EFP also showed a downward trend. The state of decoupling appeared as a recessive decoupling (2016) and a weak negative decoupling (2017 and 2018), and EFP intensity first increased and then decreased. (4) The total effect of the factors affecting the EFP of the textile industry in Shaoxing demonstrated a pulling trend, and industrial scale played a significant role in driving the EFP. The energy consumption intensity effect contributed the largest restraint. This paper fills in the gaps in the environmental regulation means and methods of pillar industrial clusters in specific regions.

Keywords: textile industry; energy footprint; decoupling; logarithmic mean Divisia index; Shaoxing

---

1. Introduction

The relationship between economic growth and environmental pressure has been vividly discussed [1–3]. Excessive greenhouse gas emissions have caused serious environmental problems, such as global warming, melting glaciers and rising sea levels. Excessive energy consumption is the cause of such problems. Approximately 63% of the gaseous radiative force that contributes to climate change is carbon [4]. To reduce energy consumption in the textile industry in China, the industry must explore energy-saving technologies and energy-efficient assessment methods [5–7]. China is the world's largest textile product manufacturer and exporter. The textile industry is a traditional pillar industry in China, an important civilian production industry and an industry that creates new international advantages [8–10]. China's cotton textile spindles account for half the world's total, while China's textile trade volume accounts for one third of the world's total [8]. However, it is a high-energy, high-emissions and high-pollution industry and one of the largest sources of greenhouse gas emissions [11,12]. The energy consumed by the textile industry is approximately 4.3% of China's total energy consumption in the manufacturing industries [13]. This makes the textile industry one of the biggest pollution emitters in China. China's Textile Industry 13th (2016–2020) Five-Year Plan highlights low-carbon development as a major strategy for economic and social development, and it is also an important way to construct an ecological civilization [14].

Shaoxing is the largest textile printing and dyeing base in the world, and its output in printing and dyeing cloths ranks first in China. Shaoxing built the China International Textile City, which is the world's largest printing and dyeing cloth trading center. In 2018, the printed fabrics output of printing and dyeing enterprises above a designated size was 16.4 billion meters in Shaoxing, thereby accounting for 56.71% of the total output of printed fabrics in Zhejiang Province and approximately one-third of the total output of printed fabrics in China [15]. The total industrial output value (TIOV) of textile enterprises above a designated size is USD 22.05 billion, accounting for 29.13% of Shaoxing's industrial output value [15]. The textile industry has made important contributions to the economic and social development of Shaoxing. However, the energy consumption of Shaoxing's textile industry in 2018 was 6.16 million tons of standard coal, thereby accounting for 36.91% of the city's industrial energy consumption [15]. The energy consumption ratio was greater than the actual output value ratio, and the unit energy output value was lower than the city's industrial average. The textile industry is a pillar industry and a typical high-pollution industry in Shaoxing. Implementing an economic energy-saving and consumption reduction model is an inevitable trend and a strategic choice in the economic development of Shaoxing's textile industry. Studying the changing rules and influencing factors of the energy footprint (EFP) of the textile industry in Shaoxing can determine the effects, experiences and deficiencies of the transformation and upgrade of typical industries and environmental regulations in different regions. Moreover, such an investigation can provide feasible countermeasures and suggestions for the high-quality development of the textile industry.

Wackernagel et al. [16] defined EFP as the forest area needed to absorb CO_2 generated by the burning of fossil energy, buffer nuclear radiation and build hydropower dams. Palamutcu [17] believed that EFP is one way to measure energy consumption. The EFP indicator, which links the production and consumption activities of the global economy with energy consumption, plays an important role in the study of sustainable development approaches [18]. EFP, as an indicator of energy production and consumption, can effectively characterize the direct and indirect energy sources required to supply end-consumer goods or services. In addition, it can provide quantitative and decision-making bases for policymakers to improve energy efficiency and optimize energy structure. Numerous scholars have studied issues related to EFP. Chen et al. [19] indicated that using the EFP index to quantify energy consumption can provide an important basis for measuring energy sustainability. Ozturk [20] estimated the energy consumption and energy costs of the Turkish textile sector and analyzed the relationship between energy use and textile product production. Hong et al. [21] studied the energy conservation status of the textile industry in Taiwan and China and provided a benchmark for measuring energy efficiency. Meanwhile, Guo et al. [22] used EFP to examine the embodied and direct energy links between global and Chinese construction industries. Previous research has also used different methods to estimate EFP based on global [23,24], national [25,26], local [27,28], industrial [29,30] and product [31,32] perspectives. Methods used include ecological footprint analysis [33], input-output analysis [34] and lifecycle assessment [35]. Studies have also measured EFP in national and global hectares, as well as in energy/functional units [36,37].

To explore the pressure of the textile industry's energy consumption reduction on the environment, this study analyzes the relationship between the energy consumption and economic growth of the Shaoxing's textile industry, based on its energy consumption data and economic development from 2005 to 2018. This study also determines the main factors affecting the relationship and provides relevant suggestions for energy-saving and emissions-reduction paths and measures the sustainable development of the Shaoxing region and textile industry. This paper fills in the gaps in the environmental regulation means and methods of pillar industrial clusters in specific regions.

2. Methods and Data Sources

2.1. EFP

EFP is an important method for measuring energy consumption. In this study, we measured EFP in units of energy and followed the definition of demand for nonrenewable energy resources. The EFP of the textile industry can be expressed as follows:

$$EFP = \sum E_j f_j, \tag{1}$$

In Equation (1), EFP is the EFP; E_j represents the consumption amount of energy j; and f_j is conversion factor from physical units to coal equivalent of energy j.

EFP intensity (EFPI) is an indicator that measures changes in energy efficiency, which is expressed as follows:

$$EFPI = \frac{EFP}{TIOV}, \tag{2}$$

In Equation (2), $EFPI$ is the EFPI, and $TIOV$ is the TIOV of textile enterprises above a designated size (billion USD).

2.2. Tapio Decoupling Indicator

Elastic analysis mainly uses elasticity to measure the degree of decoupling. Tapio improved and refined the decoupling model [38]. Decoupling is a term that refers to the status when the relationship between energy consumption and economic growth begins to break up [39]. Decoupling can be further divided into two/three/six/eight subcategories [38], and all were used in earlier studies [9,40–42]. Compared with other methods, the 8-min state classification method is more detailed and helpful for the analysis of the study. The present study analyzed the decoupling of Shaoxing's economic development and energy consumption from 2005 to 2018, based on the Tapio decoupling model, and the results were expressed by decoupling elasticity. In this regard, the decoupling function can be expressed as Equation (3):

$$E = \frac{\%\Delta EFP}{\%\Delta TIOV}, \tag{3}$$

In Equation (3), E is the decoupling elasticity index of the TIOV. Moreover, $\%\Delta EFP$ is the EFP change rate in the textile industry, and $\%\Delta TIOV$ is the TIOV change rate in the textile industry. Owing to the different values of $\%\Delta EFP$, $\%\Delta TIOV$ and E, decoupling states can be divided into eight categories, based on Tapio's elastic division. According to the order of the ideal degree of decoupling from large to small, the categories are strong, weak, recessive decoupling, expansive, recessive coupling, expansive negative, weak negative and strong negative decoupling. Strong decoupling is the most ideal state, whereas strong negative decoupling is the least ideal state. The decoupling state was determined according to the decoupling elasticity index E, and the specific division method can be seen in Table 1.

Table 1. Tapio decoupling model division method.

Decoupling State	$\%\Delta EFP$	$\%\Delta TIOV$	E
Strong decoupling	−	+	$(-\infty, 0)$
Weak decoupling	+	+	$(0, 0.8)$
Recessive decoupling	−	−	$(1.2, +\infty)$
Expansive coupling	+	+	$(0.8, 1.2)$
Recessive coupling	−	−	$(0.8, 1.2)$
Expansive negative decoupling	+	+	$(1.2, +\infty)$
Weak negative decoupling	−	−	$(0, 0.8)$
Strong negative decoupling	+	−	$(-\infty, 0)$

2.3. Logarithmic Mean Divisia Index (LMDI) Decomposition Model

The LMDI decomposition method was proposed by Ang et al. [43]. It can give a reasonable factor decomposition and has the advantage of the absence of residual and full decomposition, thereby making a model appear convincing. The LMDI factor decomposition method is mainly used to analyze the influencing factors of the total amount of economic and social indicators, such as energy consumption and pollutant emissions. In addition, it is used to determine the contribution of different influencing factors to changes in the total amount of indicators. In recent years, the LMDI has been widely applied, as well as continuously developed and optimized [9,44]. In the present study, LMDI was used to decompose the EFP of Shaoxing's textile industry. The specific formula of the decomposition model is shown in Equation (4), and the meanings of the variables are shown in Table 2.

$$EFP = \sum E_{ij} = \sum TIOV \frac{TIOV_i}{TIOV} \frac{E_i}{TIOV_i} \frac{E_{ij}}{E_i} = TIOV \cdot I_i \cdot T_i \cdot S_{ij}, \tag{4}$$

$$\begin{aligned} \Delta EFP = EFP^t - EFP^0 &= \sum TIOV^t I_i^t T_i^t S_{ij}^t - \sum TIOV^0 I_i^0 T_i^0 S_{ij}^t \\ &= \Delta EFP_{TIOV} + \Delta EFP_I + \Delta EFP_T + \Delta EFP_S, \end{aligned} \tag{5}$$

where EFP^t is the EFP of Shaoxing's textile industry in year t, and EFP^0 is the EFP of Shaoxing's textile industry in base year 0. The EFP expression formula for changes in the EFP of Shaoxing's textile industry yearly, according to an additional LMDI decomposition model, is shown in Equation (5).

Table 2. Meanings of the decomposition analysis model variables.

Variable	Meaning	Variable	Meaning
E_{ij}	the consumption amount of energy *j* in the subsector *i*	I_i	TIOV proportion of subsector *i* in textile industry, i.e., the industry structure factor
$TIOV_i$	the TIOV of subsector *i*	T_i	The energy intensity of subsector *i*
E_i	the total energy consumption in the subsector *i*	S_{ij}	The ratio of energy *j* consumption in subsector *i* to total energy consumption in the industry, i.e., the energy structure factor

In Formula (5), ΔEFP_{TIOV} represents the scale effect of the textile industry, ΔEFP_I is the industrial structure effect, ΔEFP_T denotes the energy consumption intensity effect and ΔEFP_S is the energy structure effect. The contribution values of each factor to the EFP changes of the textile industry are ΔEFP_{TIOV}, ΔEFP_I, ΔEFP_T and ΔEFP_S. The expressions of the contribution of each decomposition factor are as follows:

$$\Delta EFP_{TIOV} = \sum \frac{E_{ij}^t - E_{ij}^0}{lnE_{ij}^t - lnE_{ij}^0} ln \frac{TIOV^t}{TIOV^0} \tag{6}$$

$$\Delta EFP_I = \sum \frac{E_{ij}^t - E_{ij}^0}{lnE_{ij}^t - lnE_{ij}^0} ln \frac{I_i^t}{I_i^0} \tag{7}$$

$$\Delta EFP_T = \sum \frac{E_{ij}^t - E_{ij}^0}{lnE_{ij}^t - lnE_{ij}^0} ln \frac{T_i^t}{T_i^0} \tag{8}$$

$$\Delta EFP_S = \sum \frac{E_{ij}^t - E_{ij}^0}{lnE_{ij}^t - lnE_{ij}^0} ln \frac{S_{ij}^t}{S_{ij}^0} \tag{9}$$

2.4. Data Sources and Processing

China's textile industry is classified under three subsectors: manufacturers of textiles (MT); manufacturers of textile apparel, footwear and caps (MTAFC); and manufacturers of chemical fibers

(MCF). The TIOV and energy consumption data of enterprises above a designated size in Shaoxing's textile industry and subindustries were derived from the 2006–2019 Shaoxing statistical yearbook. The f_j data were obtained from the 2018 China energy statistical yearbook. Owing to inflation and other factors that influence constantly changing prices in economic development, the gross product of Shaoxing's textile industry was compared with its energy consumption, which was calculated by the TIOV comparable price, and 2005 was used as the base period. Nomenclature comparison table is in Appendix A.

3. Results and Discussion

3.1. Descriptive Analysis and Discussion

Figure 1 shows the EFP change trends and TIOV of Shaoxing's textile industry and subsectors from 2005 to 2018. In addition, Figure 1 shows that, from 2005 to 2018, Shaoxing's EFP demonstrated a rising and then a declining trend, which can be divided into two stages, namely the rising stage and the declining stage. The rising period was from 2005 to 2007, with an average annual growth of 4.4%. In 2007, the EFP reached the highest value of 2.11 Mt, whereas 2008–2018 was the declining period, with an average annual decline of 7.19%. In 2018, the lowest EFP was 0.45 Mt. In Figure 1, the TIOV change trends of the textile industry are also divided into two stages, namely the rising stage and the falling stage. The rising stage was from 2005 to 2015, with an average annual growth rate of 9.1%. The maximum TIOV was USD 46.97 billion in 2015. During the declining period, the annual average decline rate was 19.2% (2016–2018), in which the lowest TIOV was USD 22.01 billion in 2018.

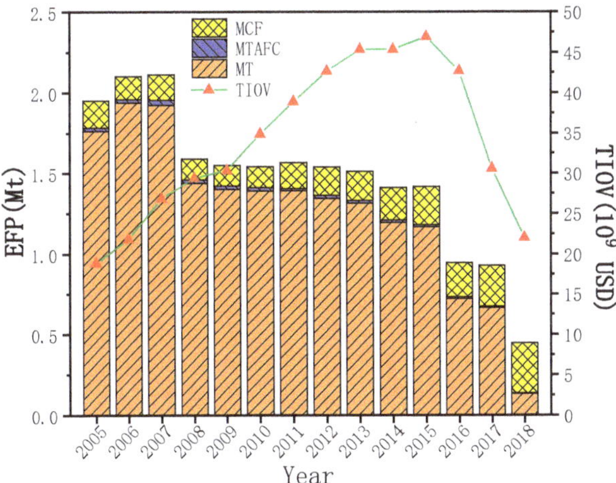

Figure 1. EFP of Shaoxing's textile industry and the subsectors and TIOV of Shaoxing's textile industry during 2005–2018.

The EFPI of Shaoxing's textile industry and subsectors from 2005 to 2018 is shown in Figure 2. In the sample period, the total EFPI of the textile industry showed a downward trend. The EFPI of the MT and MTAFC subindustries demonstrated a downward trend. Specifically, the downward trend of the MT subindustry was significant, with a decrease rate of 82.65% (2005–2016). The EFPI of the MT subindustry decreased from 0.114 (2005) to 0.020 (2016), whereas the EFPI of the MTAFC subindustry experienced a relatively small decline, from 0.00121 (2005) to 0.00022 (2018). In the MCF subindustry, EFPI first slowly declined to 0.00603 (2005–2010) and then rose to 0.01055 (2011–2018).

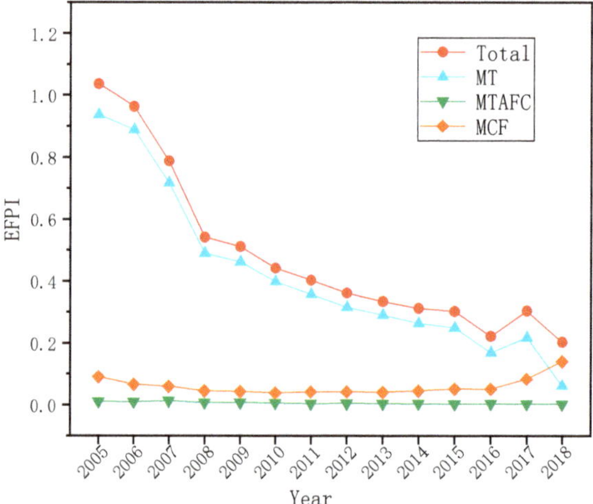

Figure 2. EFPI of Shaoxing's textile industry and the subsectors during 2005–2018.

The decoupling state between the TIOV and EFP of Shaoxing is shown in Table 3. For 13 years, from 2006 to 2018, the overall decoupling state was satisfactory. During the steady rise of the TIOV, the EFP showed a downward trend. In 10 years, from 2006 to 2015, three years of weak decoupling (2006, 2007 and 2011) was followed by seven years of strong decoupling (2008–2010 and 2012–2015). The 2016 decoupling index was 4.39, thereby indicating a recessive decoupling. In 2017 and 2018, the TIOV declined, with decline rates of 27% and 26%, respectively. Thus, the decoupling state for both years was a weak negative decoupling.

Table 3. Decoupling state between TIOV and EFP of Shaoxing.

Year	$\%\Delta EFP$	$\%\Delta TIOV$	E	Decoupling State
2006	0.08	0.17	0.45	Weak decoupling
2007	0.01	0.26	0.02	Weak decoupling
2008	−0.25	0.16	−1.55	Strong decoupling
2009	−0.03	0.03	−0.95	Strong decoupling
2010	−0.01	0.20	−0.03	Strong decoupling
2011	0.02	0.17	0.09	Weak decoupling
2012	−0.02	0.12	−0.15	Strong decoupling
2013	−0.02	0.08	−0.23	Strong decoupling
2014	−0.07	0.02	−3.55	Strong decoupling
2015	0.005	0.04	0.12	Strong decoupling
2016	−0.33	−0.08	4.39	Recessive decoupling
2017	−0.02	−0.27	0.07	Weak negative decoupling
2018	−0.04	−0.26	0.14	Weak negative decoupling

The EFP decline in Shaoxing and the increase in energy-use intensity are inseparable from the intensity of environmental protection in China and Shaoxing. The sample period can be divided into three periods, according to Figures 1 and 2 and Table 3.

The first period (2005–2007): In the years after China's accession to the World Trade Organization in 2001, Shaoxing mainly focused on the rapid development of the textile industry. Shaoxing, as a globally important textile printing and dyeing product production base, was driven by international and Chinese domestic demands. As a result of the increase in production, international trade in textile products was prosperous and TIOV increased. However, the production and supply of products consumed considerable energy, thereby leading to serious environmental pollution problems. The amount of energy consumed per output value unit (EFPI) decreased, which meant that the increase in the TIOV was greater than the increase in the EFP, and energy efficiency improved. Therefore, the decoupling state at this period was a weak decoupling. Thus, the coupling relationship between energy consumption and economic growth was not completely broken.

The second period (2008–2015): After energy consumption caused environmental pollution, China vigorously promoted the construction of an ecological civilization. Furthermore, various parties in society adopted numerous options and measures, including implementing environmental regulations, adopting market-based trading methods and conducting technological innovations. Entrepreneurs' self-environmental and public environmental awareness also improved. Firstly, environmental regulations were created during the period, specifically, in the textile industry of Zhejiang Province and Shaoxing Municipality, including laws, regulations [45,46], notices, industry standards and pollution permits. Secondly, market-oriented means were developed through the implementation of emissions trading systems, pollution fees and emissions taxes. Such measures could effectively control energy consumption and total emissions to reduce pollution emissions. Thirdly, regarding the technological innovation of textile enterprises, pressure on environmental protection forced the technological transformation of enterprises. At the same time, the government provided subsidies for technological transformation, thereby effectively improving the technological efficiency of the textile industry. Fourthly, entrepreneurs' awareness of environmental protection and social responsibility was enhanced, investment in environmental protection increased and the 'changing coal into natural gas' [47] cooperation was consciously implemented to reduce carbon emissions; the natural gas is a secondary energy source. Such factors led to the strong decoupling of energy consumption from economic growth during this period. Whilst TIOV increased, energy consumption decreased, and EFPI decreased further.

The third period (2016–2018): In January 2016, Shaoxing implemented 'Challenging by Showing Sword' to promote the upgrade of the textile printing and dyeing industry. 'Challenging by Showing Sword' referred to the centralized rectification of textile printing and dyeing enterprises that generated large amounts of pollution and had hidden safety hazards and chaotic management. Specifically, this operation involved the rectification of several textile printing and dyeing enterprises with low production capacities, small scales and unregulated pollution discharges. Enterprises affected the image, product quality and price order of the textile printing and dyeing industry in Shaoxing and restricted the overall improvement and development of the printing and dyeing industry to a certain extent. During this operation, the production of an initial batch of 74 textile printing and dyeing companies was suspended, owing to rectification, including 10 in the Paojiang Industrial Park and 64 in the Keqiao Industrial Park. EFPI showed an increasing and then decreasing trend. Energy consumption and environmental pollution problems were effectively controlled during this second period, but a hysteresis effect was generated, thereby creating a situation in which the TIOV no longer increased but declined sharply. Therefore, the states of decoupling during this period did not seem ideal, that is, a recessive decoupling, a weak negative decoupling and a weak negative decoupling. The reason for the less-than-ideal situations was that, after enterprises dispersed from the relocation to the Shaoxing Binhai Industrial Park, several enterprises with low-end production capacities were shut down, especially those with poor product quality, large energy consumption, large chemical consumption and weak production capacities. In the industrial agglomeration zone, textile printing and dyeing enterprises uniformly managed and discharged sewage and implemented strict emissions standards. Thus, enterprises considerably improved in terms of management, innovation, equipment, technology, environment, entrepreneurship and labor quality. The core competitiveness and industry voice of textile printing and dyeing enterprises in Shaoxing were also strengthened.

3.2. LMDI Decomposition Analysis and Discussion

The EFP of Shaoxing's textile industry in 2006 and 2018 was decomposed by the LMDI factor decomposition method. The factor decomposition trend is shown in Figure 3. The four main factors included the industrial scale effect, the industrial structure effect, the energy consumption intensity effect and the energy structure effect. In addition, total effect was calculated. From 2006 to 2018, the total EFP effect of the textile industry showed a pulling trend. However, the pulling range dropped in waves, with a maximum pulling effect of 1.57 Mt (2007) and a minimum of 0.12 Mt (2018). Among the four factors, the cumulative effect of the industrial scale was a positive factor, which exerted the largest pulling effect on the EFP. The cumulative pulling effect was 37.61 Mt during the sample period, which was higher than the total inhibitory effect of the other three factors (i.e., 25.60 Mt). Thus, the total effect was positive. From 2006 to 2015, the scale effect of the industry increased and then decreased sharply in the next three years, thereby creating an inverted 'U' shape, with a peak value of 4.50 Mt (2015). The energy consumption intensity effect provided the maximum inhibitory effect, thereby offsetting the industrial scale pull effect, with a total offset of 20.80 Mt. The effect of energy consumption intensity decreased from 2006 to 2014 then increased yearly, thereby pulling EFP growth in 2006 and 2018 and creating a 'U' shape, with a lowest point of −3.05 Mt (2014). The effect of industrial structure and energy structure on EFP was small, with a cumulative inhibition of 4.80 Mt. After 2011, the energy structure effect increased, showed a pull effect from 2011 to 2014 and then declined after 2014, thereby demonstrating an inhibition effect. The industrial structure effects were inhibitory effects during the sample period (except for 2009), and the inhibitory effect was enhanced after 2011, with a cumulative inhibition of 2.95 Mt.

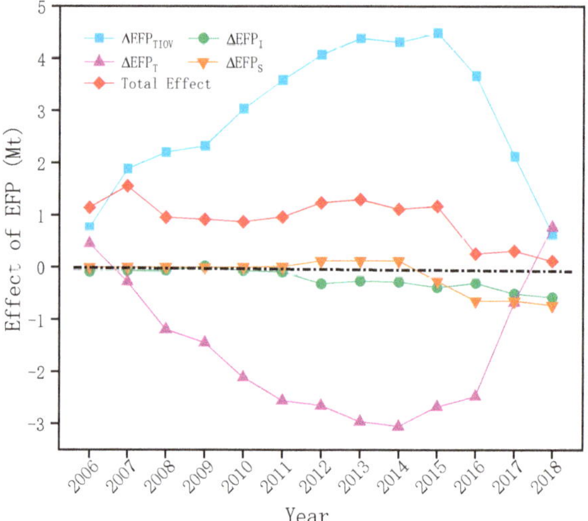

Figure 3. EFP effect decomposition of Shaoxing's textile industry during 2006–2018.

From the perspective of the influencing factors of the relationship between EFP and economic growth, the cumulative effect of industrial scale factors had the largest pulling effect on EFP. The MT sector made up the largest proportion of the textile subsectors of Shaoxing. Compared with the other two subsectors, the MT sector is a typical high energy-consuming sector and the most energy-consuming and polluting sector in the printing and dyeing industry. A large EFP is generated during dyeing, printing and finishing processes. Printing and dyeing subsectors have a large base and high energy consumption in the MT sector, and the potential for energy conservation, consumption reduction and emissions reduction is relatively small. In this investigation, we found that most of the printing and dyeing enterprises in Shaoxing were mainly based on sample processing. Moreover, most followed imitations in terms of process technology, variety development and management, with few independent brands and poor R&D and innovation capabilities. The online parameter detection and online control technology of dyeing and finishing equipment, as well as the development and manufacturing accuracy of new equipment and supporting parts, energy saving and environmental protection, lag behind advanced global levels. Therefore, optimizing the printing and dyeing process and improving industrial efficiency are urgent for the Shaoxing's textile industry to achieve high-quality development.

The energy structure effect is a crucial factor that restrains the growth of energy consumption. Figure 4 shows the energy structure of the textile industry in Shaoxing from 2005 to 2018. Figure 4 indicates that raw coal, heat and electricity are the main sources of energy consumption in Shaoxing's textile industry. With the Shaoxing government's introduction of the 'changing coal into natural gas' policy and textile enterprises' updating of advanced technology and equipment, the use of electricity and heat increased significantly from 2005 to 2018, and the growth rate of heat was greater than that of electricity. Raw coal that produced a considerable amount of pollution decreased significantly during the sample period. Natural gas is the cleanest fossil fuel, with relatively low carbon intensity than coal in most end uses [48,49]. Heat produced by natural gas is a clean energy source that create CO_2 emissions less than raw coal in Shaoxing's textile industry [49]. In general, clean energy accounted for an increasing proportion of the energy structure of the Shaoxing's textile industry. Furthermore, the consumption of electricity and heat continued to cause EFP consumption indirectly. Therefore, Shaoxing should further optimize energy structure and gradually increase the proportion of clean energy. Such optimization includes increasing the proportion of secondary energy converted from clean energy, which in turn will increase the significance of the energy conservation and environmental protection effect.

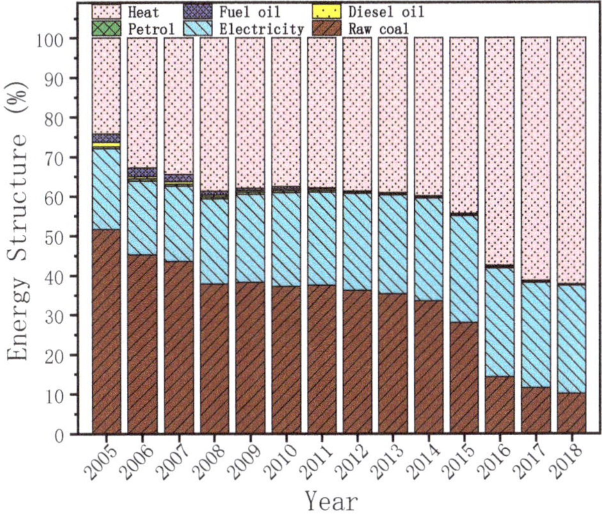

Figure 4. The energy structure of Shaoxing's textile industry during 2005–2018.

4. Conclusions

Energy consumption is a significant factor that affects the sustainable development of Shaoxing's textile industry. A detailed analysis of energy consumption in Shaoxing's textile industry can provide essential information for the establishment of energy policies. This study calculates the EFP of the textile industry in Shaoxing, from 2005 to 2018, and analyzes the relationship between the energy consumption and economic growth in the textile industry, as well as the four factors affecting the EFP. This study presents the following results:

(1) From 2005 to 2018, the EFP of Shaoxing first increased then decreased, with an average annual growth rate of 4.4% during the growth period (2005–2007) and an average annual decline of 7.19% during the decline period (2008–2018). The EFPI decline trend of the textile industry and MT

subsector was roughly the same, with the decline trend of the MT subsector as the largest among the three subsectors. However, the change trends of the MTAFC and MCF subsectors were not obvious. Shaoxing's textile enterprises should accelerate the development and production of ecological and functional textiles and promote environmentally friendly, energy-saving and clean production printing and dyeing technologies. Innovation funds should be set up by Shaoxing's government and social investors, to alleviate pressure in the capital of small- and medium-sized enterprises

(2) For 13 years, from 2006 to 2018, the overall decoupling state was satisfactory. In the 10 years from 2006 to 2015, three years of weak decoupling and was followed by seven years of strong decoupling. In 2016, the state was a recessive decoupling. In 2017 and 2018, TIOV declined, and both years showed a weak negative decoupling. Energy consumption restriction policies and the popularization of energy-saving technologies promote the decoupling of energy consumption from economic growth, diminishing the gap between promoting factors and inhibiting factors. Shaoxing's government should eliminate obsolete process equipment and shut down/close printing and dying enterprises that consume considerable amounts of energy and generate high pollution, as well as wastewater treatment facilities that cannot meet requirements.

(3) The total energy footprint effect of the textile industry showed a pulling trend from 2006 to 2018 but dropped in waves. The cumulative effect of the industrial scale was a positive factor, which played a significant role in driving the EFP. Technological innovation and cutting-edge technology applications in the textile industry of Shaoxing are necessary in order to to reduce the energy consumption. These will also improve the industrial productivity. The cumulative effects of the other three factors were negative, and the sum of the inhibitory effect was smaller than that of the industrial scale effect, among which the effect of energy consumption intensity contributed the largest inhibitory effect. Environmental policies and laws should be strengthened to reduce the proportion of the energy intensive sector and promote high value-added textile products manufacturing with less energy intensity.

Both TIOV and EFP had been rapidly decreasing for Shaoxing's textile industry after 2015. Consequently, structural unemployment appeared, so many skilled textile workers were laid off. Supply and demand of textile products may be unbalanced in the future. Such problems should be further discussed in the future studies.

Author Contributions: Conceptualization, Y.L. and Y.Y.; methodology, L.Z.; validation, X.W., X.C., Y.C. and L.Z.; formal analysis, X.W.; investigation, X.C.; resources, Y.C.; data curation, Y.Y.; writing—original draft preparation, X.C., Y.C. and L.Z.; writing—review and editing, Y.L.; visualization, X.W.; supervision, Y.Y.; project administration, Y.L.; funding acquisition, X.W., Y.L. and Y.Y. All authors have read and agreed to the published version of the manuscript.

Funding: This research was funded by the Project of Philosophy and Social Sciences in Zhejiang Province, China (19NDJC235YB), Key projects of the national social science fund (19AZD004), General Project of Humanities and Social Sciences Research of the Ministry of Education of China (19YJCZH092), Key Project of Zhejiang Ecological Civilization Research Center of Zhejiang Provincial Key Research Base of Philosophy and Social Sciences and K. C. Wong Magna Fund in Ningbo University.

Acknowledgments: We are very grateful for the support of the School of Economics and Management/Fashion School of Zhejiang Sci-Tech University, and Zhejiang Silk and Fashion Culture Research Center for this study.

Conflicts of Interest: The authors declare no conflict of interest.

Appendix A

Table A1. Nomenclature comparison table.

Abbreviation	Detailed Name
EFP	Energy footprint
EFPI	Energy footprint intensity
TIOV	Total industrial output value
LMDI	Logarithmic mean Divisia index
MT	Manufacturer of textiles
MTAFC	Manufacturer of textile apparel, footwear and caps
MCF	Manufacturer of chemical fibers

References

1. Holdren, J.P. Science and Technology for Sustainable Well-being. *Science* **2008**, *319*, 424–434. [CrossRef]
2. Magazzion, C. The Relationship among Economic Growth, CO_2 Emissions and Energy Use in the APEC Countries: A Panel VAR Approach. *Environ. Syst. Decis.* **2017**, *37*, 353–366. [CrossRef]
3. Li, Y.; Luo, Y.; Wang, Y.; Wang, L.; Shen, M. Decomposing the decoupling of water consumption and economic growth in China's textile industry. *Sustainability* **2017**, *9*, 412. [CrossRef]
4. Wang, Q.; Li, R.; Jiang, R. Decoupling and Decomposition Analysis of Carbon Emissions from Industry: A Case Study from China. *Sustainability* **2016**, *8*, 1059. [CrossRef]
5. Shen, Y.; Lisa, A.A. Local environmental governance innovation in China: Staging 'triangular dialogues' for industrial air pollution control. *J. Chin. Gov.* **2018**, *3*, 351–369. [CrossRef]
6. Guan, T.; Delman, J. Energy policy design and China's local climate governance: Energy efficiency and renewable energy policies in Hangzhou. *J. Chin. Gov.* **2017**, *2*, 68–90. [CrossRef]
7. Li, Y.; Ding, L.; Yang, Y. Can the Introduction of an Environment Target Assessment Policy Improve the TFP of Textile Enterprises? A Quasi-Natural Experiment Based on the Huai River in China. *Sustainability* **2020**, *12*, 1696. [CrossRef]
8. Zhao, H.; Lin, B. Resources allocation and more efficient use of energy in China's textile industry. *Energy* **2019**, *185*, 111–120. [CrossRef]
9. Wang, L.; Li, Y.; He, W. The energy footprint of China's textile industry: Perspectives from decoupling and decomposition analysis. *Energies* **2017**, *10*, 1461. [CrossRef]
10. Li, Y.; Shen, J.; Lu, L.; Luo, Y.; Wang, L.; Shen, M. Water environment stress, rebound effect, and economic growth of China's textile industry. *PeerJ* **2018**, *6*, e5112. [CrossRef]
11. Zhao, H.; Lin, B. Impact of foreign trade on energy efficiency in China's textile industry. *J. Clean. Prod.* **2020**, *245*, 118878. [CrossRef]
12. Li, Y.; Wang, Y. Double decoupling effectiveness of water consumption and wastewater discharge in China's textile industry based on water footprint theory. *PeerJ* **2019**, *7*, e6937. [CrossRef]
13. China Statistics Press. *China Statistical Yearbook on Science and Technology*; China Statistics Press: Beijing, China, 1981–2013. (In Chinese)
14. Notice of the Ministry of Industry and Information Technology on Distributing the Development Plan of the Textile Industry (2016–2020) MIIT Regulation [2016] No. 305. Available online: http://www.miit.gov.cn/n1146295/n1652858/n1652930/n3757019/c5267251/content.html (accessed on 28 February 2020). (In Chinese)
15. Shaoxing Statistical Yearbook 2019. Available online: http://tjj.sx.gov.cn/art/2019/12/31/art_1489207_41684038.html (accessed on 13 February 2020). (In Chinese)
16. Wackernage, M.; Lewan, L.; Hansson, C.B. Evaluating the use of natural capital with the ecological footprint Applications in Sweden and Subregions. *AMBIO* **1999**, *28*, 604–612.
17. Palamutcu, S. *Energy Footprints in the Textile Industry*; Woodhead Publishing: Cambridge, UK, 2015; pp. 31–61.
18. Lan, J.; Malik, A.; Lenzen, M.; McBain, D.; Kanemoto, K. A structural decomposition analysis of global energy footprints. *Appl. Energy* **2016**, *163*, 436–451. [CrossRef]
19. Chen, S.Q.; Kharrazi, A.; Liang, S.; Fath, B.D.; Lenzen, M.; Yan, J.Y. Advanced approaches and applications of energy footprints toward the promotion of global sustainability. *Appl. Energy* **2020**, *261*, 114415. [CrossRef]

20. Ozturk, H.K. Energy usage and cost in textile industry: A case study for Turkey. *Energy* **2005**, *30*, 2424–2446. [CrossRef]
21. Hong, G.B.; Su, T.L.; Lee, J.D.; Hsu, T.C.; Chen, H.W. Energy conservation potential in Taiwanese textile industry. *Energy Policy* **2010**, *38*, 7048–7053. [CrossRef]
22. Guo, S.; Zheng, S.; Hu, Y.; Hong, J.; Wu, X.; Tang, M. Embodied energy use in the global construction industry. *Appl. Energy* **2019**, *256*, 113838. [CrossRef]
23. Chen, G.Q.; Wu, X.F. Energy overview for globalized world economy: Source, supply chain and sink. *Renew. Sustain. Energy Rev.* **2017**, *69*, 735–749. [CrossRef]
24. Chen, Z.M.; Chen, G.Q. An overview of energy consumption of the globalized world economy. *Energy Policy* **2011**, *39*, 5920–5928. [CrossRef]
25. Chen, B.; Chen, G.Q.; Yang, Z.F.; Jiang, M.M. Ecological footprint accounting for energy and resource in China. *Energy Policy* **2007**, *35*, 1599–1609. [CrossRef]
26. Zhang, B.; Qiao, H.; Chen, Z.M.; Chen, B. Growth in embodied energy transfers via China's domestic trade: Evidence from multi-regional input-output analysis. *Appl. Energy* **2016**, *184*, 1093–1105. [CrossRef]
27. Zhang, B.; Chen, Z.M.; Xia, X.H.; Xu, X.Y.; Chen, Y.B. The impact of domestic trade on China's regional energy uses: A multi-regional input-output modeling. *Energy Policy* **2013**, *63*, 1169–1181. [CrossRef]
28. Wilson, J.; Tyedmers, P.; Grant, J. Measuring environmental impact at the neighbourhood level. *J. Environ. Plan. Manag.* **2013**, *56*, 42–60. [CrossRef]
29. Bordigoni, M.; Hita, A.; Blanc, G.L. Role of embodied energy in the European manufacturing industry: Application to short-term impacts of a carbon tax. *Energy Policy* **2012**, *43*, 335–350. [CrossRef]
30. Crawford, R.H. Validation of the use of input-output data for embodied energy analysis of the australian construction industry. *J. Constr. Res.* **2005**, *6*, 71–90. [CrossRef]
31. Zhao, Y.; Onat, N.C.; Kucukvar, M.; Tatari, O. Carbon and energy footprints of electric delivery trucks: A hybrid multi-regional input-output life cycle assessment. *Transp. Res. Transp. Environ.* **2016**, *47*, 195–207. [CrossRef]
32. Lenzen, M.; Dey, C. Truncation error in embodied energy analyses of basic iron and steel products. *Energy* **2000**, *25*, 577–585. [CrossRef]
33. Wiedmann, T. A first empirical comparison of energy Footprints embodied in trade-MRIO versus PLUM. *Ecol. Econ.* **2009**, *68*, 1975–1990. [CrossRef]
34. Wang, H.; Wang, G.; Qi, J.; Schandl, H.; Li, Y.; Feng, C.; Liang, S. Scarcity-weighted fossil fuel footprint of China at the provincial level. *Appl. Energy* **2020**, *258*, 114081. [CrossRef]
35. Chang, Y.; Ries, R.J.; Wang, Y. The embodied energy and environmental emissions of construction projects in China: An economic input-output LCA model. *Energy Policy* **2010**, *38*, 6597–6603. [CrossRef]
36. Cucek, L.; Klemes, J.J.; Kravanja, Z. A review of footprint analysis tools for monitoring impacts on sustainability. *J. Clean. Prod.* **2012**, *34*, 9–20. [CrossRef]
37. Kenourgios, D.; Drakonaki, E.; Dimitriou, D. ECB's unconventional monetary policy and cross-financial-market correlation dynamics. *N. Am. J. Econ. Financ.* **2019**, *50*, 101045. [CrossRef]
38. Tapio, P. Towards a theory of decoupling: Degrees of decoupling in the EU and the case of road traffic in Finland between 1970 and 2001. *Transp. Policy* **2005**, *12*, 137–151. [CrossRef]
39. Hao, Y.; Huang, Z.; Wu, H. Do Carbon Emissions and Economic Growth Decouple in China? An Empirical Analysis Based on Provincial Panel Data. *Energies* **2019**, *12*, 2411. [CrossRef]
40. Zhang, Y.; Zou, X.; Xu, C.; Yang, Q. Decoupling Greenhouse Gas Emissions from Crop Production: A Case Study in the Heilongjiang Land Reclamation Area, China. *Energies* **2018**, *11*, 1480. [CrossRef]
41. Roman, R.; Cansino, J.M.; Botia, C. How far is Colombia from Decoupling? Two-level decomposition analysis of energy consumption changes. *Energy* **2018**, *148*, 687–700. [CrossRef]
42. Li, Y.; Lu, L.; Tan, Y.; Wang, L.; Shen, M. Decoupling water consumption and environmental impact on textile industry by using water footprint method: A case study in China. *Water* **2017**, *9*, 124. [CrossRef]
43. Ang, B.W. The LMDI approach to decomposition analysis: A practical guide. *Energy Policy* **2005**, *33*, 867–871. [CrossRef]
44. Xia, C.; Li, Y.; Ye, Y.; Shi, Z.; Liu, J. Decomposed driving factors of carbon emissions and scenario analyses of low-carbon transformation in 2020 and 2030 for Zhejiang Province. *Energies* **2017**, *10*, 1747. [CrossRef]

45. The People's Government of Zhejiang Province during the '12th Five-Year Plan' Period of Heavy Pollution and High Energy Consumption, the Guiding Opinions on Deepening the Rectification of the Industry and Promoting Improvement (Zhe Zheng Fa [2011] No. 107). Available online: http://www.zj.gov.cn/art/2012/7/14/art_1582424_26042073.html (accessed on 13 February 2020). (In Chinese)
46. Implementation Plan for the Transformation and Upgrade of the Printing and Dyeing Industry in Shaoxing City (Shao Zheng Ban Fa [2013] No. 106). Available online: http://www.sx.gov.cn/art/2013/7/16/art_1467567_160160.html (accessed on 13 February 2020). (In Chinese)
47. Notice of the Shaoxing Municipal People's Government Office on Accelerating the Coal-to-Natural Gas Work of Printing and Dyeing Enterprises (Shao Zheng Ban Fa (2012) No. 158). Available online: http://www.sx.gov.cn/art/2013/2/19/art_1463812_18069826.html (accessed on 13 February 2020). (In Chinese)
48. Liu, Z.; Guan, D.; Douglas, C.B.; Zhang, Q.; He, K.; Liu, J. A low-carbon road map for China. *Nature* **2013**, *500*, 143–145. [CrossRef] [PubMed]
49. Qin, Y.; Höglund-Isaksson, L.; Byers, E.; Feng, K.; Wagner, F.; Peng, W.; Mauzerall, D.L. Air quality–carbon–water synergies and trade-offs in China's natural gas industry. *Nat. Sustain.* **2018**, *1*, 505–511. [CrossRef]

Article

Infrastructuring the Circular Economy

André Nogueira [1,*], Weslynne Ashton [2], Carlos Teixeira [3], Elizabeth Lyon [4] and Jonathan Pereira [4]

1 Harvard T.H. Chan School of Public Health, Harvard University, 677 Huntington Avenue, Boston, MA 02115, USA
2 Stuart School of Business, Illinois Institute of Technology, 10 West 35th Street, Chicago, IL 60616, USA; washton@iit.edu
3 Institute of Design, Illinois Institute of Technology, 10 West 35th Street, Chicago, IL 60616, USA; carlos@id.iit.edu
4 Plant Chicago, 4459 South Marshfield Avenue, Chicago, IL 60609, USA; elizabeth@plantchicago.org (E.L.); jonathan@plantchicago.org (J.P.)
* Correspondence: anogueira@hsph.harvard.edu

Received: 18 February 2020; Accepted: 30 March 2020; Published: 8 April 2020

Abstract: The circular economy (CE), and its focus on the cycling and regeneration of resources, necessitates both a reconfiguration of existing infrastructures and the creation of new infrastructures to facilitate these flows. In urban settings, CE is being realized at multiple levels, from within individual organizations to across peri-urban landscapes. While most attention in CE research and practice focuses on organizations, the scale and impact of many such efforts are limited because they fail to account for the diversity of resources, needs, and power structures across cities, consequently missing opportunities for adopting a more effective and inclusive CE. Reconfiguring hard infrastructures is necessary for material resource cycling, but intervening in soft infrastructures is also needed to enable more inclusive decision-making processes to activate these flows. Utilizing participatory action research methods at the intersection of industrial ecology and design, we developed a new framework and a model for considering and allocating the variety of resources that organizations utilize when creating value for themselves, society, and the planet. We use design prototyping methods to synthesize distributed knowledge and co-create hard and soft infrastructures in a multi-level case study focused on urban food producers and farmers markets from the City of Chicago. We discuss generalized lessons for "infrastructuring" the circular economy to bridge niche-level successes with larger system-level changes in cities.

Keywords: circular economy; design; industrial ecology; infrastructure; participatory action research; socio-ecological-technical systems

---

1. Introduction

The circular economy (CE) has captured the imagination of civil society and corporate actors as a framework for realizing the elusive goals of sustainability [1]. It has presented a tangible way to contribute to the sustainability of resource flows and allocation in contemporary production and consumption systems, by focusing on improving the effectiveness of current management practices in organizations through extending the useful life of products, and sharing, recycling, and regenerating resources [2].

Cities face urgent needs to expand and advance CE adoption and sustainability-oriented practices due to growing concerns over climate change, environmental pollution, and the inequitable distribution and allocation of resources in the linear (take-make-waste) economy [3,4]. They present unique opportunities for CE interventions as they are territories where human populations are concentrated, and where multiple natural (ecological) and man-made (social and technical) systems intersect, diverse human and non-human agents interact, and different types of resources are created, transformed, circulated, used, and wasted [5]. Cities are complex socio-ecological-technical systems (SETS) with nested subsystems such as neighborhoods, organizations, and infrastructure networks, and are themselves part of larger SETS such as states and nations.

In cities, social, ecological, and technical systems are deeply intertwined since the parts of one subsystem belong to and dynamically affect the others. Urban infrastructures are key elements that integrate these systems and are thus responsible for the interconnectivity and interdependency of resource flows and allocation across their levels; for example, policy and strategies at the macro-level, organizations and operations at the meso-level, and patterns of daily life and related tactics at the micro-level [6].

1.1. Urban Infrastructures

Modern urban infrastructures have become local nodes of global operations, and the means through which individuals and organizations can (or cannot) access and mobilize resources that traverse local, regional, and global sub-systems. When modern urban infrastructures were built, they did not anticipate the multitude of resources that would flow through them, the speed at which these flows would travel, and the complex inter-linked ramifications they would promote across social, ecological, and technical subsystems. Most of them were predicated on the goals of economic growth and progress, designed through the technical lenses of health and safety, and informed by principles of functionality and efficiency [7]. Yet, they failed to account for the negative and positive feedback created as resources were mobilized at faster rates than could be assimilated within urban and larger-scale ecosystems across the globe [8,9]. Intervening towards sustainability through the narrow focus of circular flows of resources among organizations ignores the many other flows that these organizations are trafficking (e.g., political agendas, cultural preferences), which have an equal or greater impact on how well-being is experienced locally [10].

Urban food distribution infrastructure in the United States (US), for example, was designed to aggregate commodities produced on large-scale, industrial farms, and distribute them to populations across the nation [11]. While these infrastructures enabled economies of scale and reduced the prices paid by consumers, they supported the generation of significant amounts of waste and accelerated environmental degradation along the entire value chain at an unprecedented speed. They also privileged economic growth and profitability of large producers over the development of local economies [11–14]. Thus, small producers, particularly ethnic minority-owned businesses in urban and peri-urban areas in the US, face challenges with access to capital and formally entering the food supply chain [12]. Environmental challenges promoted by the use of modern infrastructures include high energy and water use during production, large carbon footprints in transportation, and large volumes of waste throughout its stages [14,15]. Equally challenging are aspects related to social and economic dimensions because low income and ethnic minority populations often face disproportionate health burdens as they lack access to healthy, nutritious food, due to low availability and affordability of such food in their neighborhoods [16,17]. The food distribution infrastructure is representative of how modern infrastructures became the pathways for the circulation of many types of resources (e.g., money, knowledge, power, etc.) that were disconnected from the dynamic needs and interactions of local populations. This food production and distribution scenario reflects the complex challenges found in many urban SETS, in which multiple subsystems and variables interact, producing both desirable and undesirable emergent outcomes and feedback among the system components that can become entrenched and difficult to disrupt and change [18].

Modern infrastructures were conceptualized and built to last decades, if not centuries. They consolidated and perpetuated unsustainable and inequitable patterns of flows of resources within and across technical systems, directly influencing the well-being of people, organizations, and the natural environment (e.g., public safety policies, mobility systems, remediation processes, etc.). Over time, knowledge about how to intervene in infrastructures evolved with a distinction between two dimensions. Hard infrastructures relate to tangible and material aspects. It usually results from engineering and natural science design efforts and includes mostly technological elements, such as products and their mechanisms of operation. Soft infrastructures relate to institutions, intangible aspects, and social behavior. They are centered on the exploration of human interactions, services, and networks, and therefore, consider multiple perspectives towards unfolding new dynamics in systems [19,20].

The discussion between hard and soft dimensions reflects the paradoxical and unruly nature of infrastructures. According to Larkin, "the duality of infrastructures indicates that when they operate systemically they cannot be theorized in terms of the object alone" [21] (p. 329). On the one hand, they symbolically represent the idea of the commons, and the possibility of access to resources and assets that facilitate everyday life. On the other hand, they implicitly condition everyday life through their design, consequently reinforcing established power dynamics in the circulation and allocation of resources. Hard and soft infrastructures connect social networks by providing access to different types of resources and shaping the context for how people can or cannot work, learn, play, and live with others. As they are considered mature elements upon which activities of daily life and environmental performance fundamentally depend [22], the existing infrastructures shaping production and consumption systems are usually perceived as given and unchangeable. This is one of the reasons why contemporary CE practices are constrained by them.

1.2. Urban Infrastructures and CE

CE initiatives at the city-scale can benefit from theoretical debates around the role of infrastructure in shaping social, ecological, and technical dynamics [23–26]. In their studies, Star and Ruhleder recognized that properly working infrastructures are formed considering a set of standards and protocols of both its soft and hard dimensions [27]. According to the authors, once built, infrastructures carry a system of offerings (e.g., people, objects, environments, messages, and services) and affordances that standardize the circulation and allocation of resources, as well as how the infrastructure is used. Instead of approaching infrastructure as an element "which runs underneath actual structures", they suggested individuals and organizations recognize them as relational elements "upon which something else rides, or works, a platform of sorts" [27] (p. 151). Such an approach is particularly useful for CE initiatives happening within cities, where new technologies and new dynamics of daily life are rapidly changing and the fairly stable, technical elements of the 20th-century infrastructure are posing significant barriers to progress towards overcoming 21st-century sustainability and equity challenges.

Urban change agents are searching for creative alternatives for CE practices, business models, and new offerings that have higher "fitness" between local socio–ecological dynamics and the technical capabilities necessary to actualize a CE [28,29]. These include citizen-led material reuse and recycling centers, such as La Recyclerie in Paris, France [30] and Recycle Here in Detroit, Michigan [31], as well as local government-led CE initiatives, such as in Amsterdam, Netherlands [32] and Charlotte, North Carolina [33]. However, many CE practices within urban environments often remain novelties at the meso- or niche-level and are unable to scale as they attempt to activate and mobilize multiple resources through pathways that counter the linear logic underlying the design of these infrastructures.

Furthermore, many individuals and organizations who are exploring large-scale CE urban interventions operate within their own traditional disciplinary silos, leverage existing networks of partners, or do not involve residents and local organizations in their processes. As a result, these interventions tend to be led, funded, and implemented by a small set of agents that lack expertise about the dynamics of the daily life of residents and practices of local organizations [34,35]. Without expanding their scope of closing loops in material resources from existing organizational practices, CE interventions at the city-level will continue to fall short in understanding how urban dynamics are shaped by a much more complex web of interconnected infrastructures responsible for allocating and mobilizing social, ecological, and technical flows of resources.

To effectively realize the CE in an urban context, micro-level changes must be able to leverage existing infrastructure or be robust enough to transform them in order to achieve lasting structural

Tools, frameworks, and methods from the three fields (IE, ST, D) were used as artifacts to facilitate and mediate engagements during prototyping activities for both discoveries of system dynamics and refinement of concepts, workshops and focus groups, and other hands-on activities of PAR. Material and energy flow analysis (MEFA) and life cycle assessment (LCA), as well as network analysis, were used to measure and map the flow and interactions of materials, money, and relationships among actors. Conventionally, MEFA is applied to track the inputs, outputs, and transformation of specific material and energy resources through a system of interest, and uses existing business practices, databases, inventories, and surveys as its primary sources for data [44]. In parallel to quantifying the materials consumed and produced by different actors, researchers leveraged IE tools to explore the potential for material reuse and cycling (industrial symbiosis) with each one of the businesses. Combined, these activities allowed for identification of common barriers across organizations concerning data gathering and sharing. Likewise, traditional network analysis maps different types of relationships among actors and quantifies their correlations. However, since researchers involved different business owners and staff in their mapping activities, different perceptions about business interactions and the multiple types of values being exchanged between them were also surfaced.

Systems dynamics modeling helped to surface key variables influencing the interaction among agents and discuss patterns of challenges underlying engaging in CE practices; for example, the nature of the relationships between organizations considering business to business, informal trading of different types of resources, and social interactions [45]. By mapping causal relationships and feedback mechanisms among these variables, researchers and partners explored leverage points of interventions, considering how changes in specific components could affect the whole [46].

Design methods and frameworks from the whole view model [47] were used as a structure for brokering different types of knowledge distributed among diverse agents throughout all phases [48]. These include, but are not limited to, performing field and user observations, and facilitating activities with different tenants and their staff for describing the current state of their organizations and their offerings, as well as to speculate alternative futures. Having a rigorous and pliant structure to perform various activities allowed the research team to better understand the patterns of strategic and operational challenges across businesses located in the facility, as well as patterns of daily life of those working in the building.

Combined, MEFA, LCA, network and systems dynamics maps, and the whole view model helped researchers to co-create a more holistic baseline with multiple agents that resulted not only in better understanding the parts of the SETS that were embedded, but also the various relationships among them through the flow and allocation of different types of resources.

3. Results

This research project can be described considering three central foci: (1) quantifying material and energy flows and circular economy potential (facility-level); (2) activating and mobilizing agents for advancing circular economy practices (facility-level); and (3) scaling niche-level CE success to city-level impact (city-level). Although the center of gravity expanded, researchers continued to perform all activities related to previous foci throughout the project. Thus, the results in both levels were generated iteratively alongside framework and model development. That is, the outputs of prior activities led to new questions, for which new frameworks and models were identified or created, which led to new outputs, expanding the scope of the project for the next iteration and expansion. Figure 2 presents an overview of the iterative research processes.

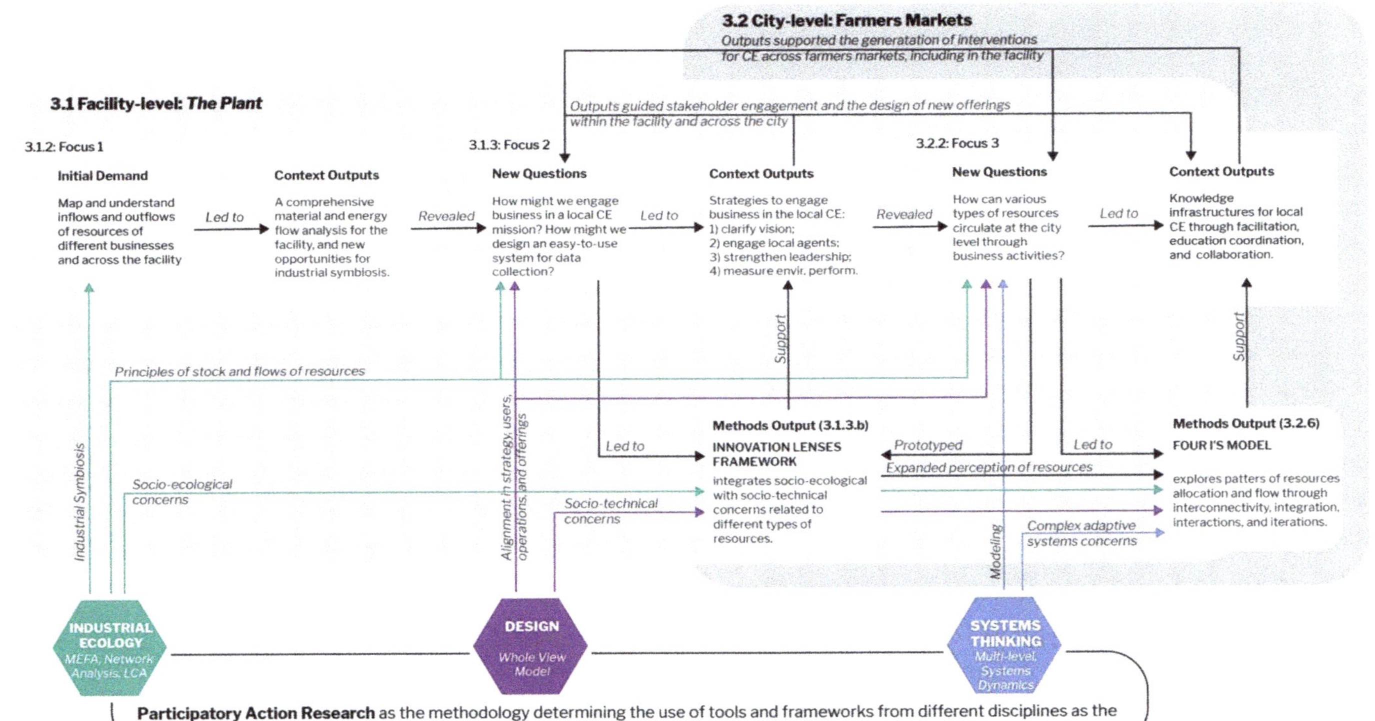

Figure 2. Iterative development of methods and results across facility and city-level projects. Abbreviations: CE, circular economy; MEFA, material and energy flow analysis; LCA, life cycle assessment.

3.1. Facility-Level: The Plant

3.1.1. Contextual Background and Problem Framing

The Plant, located in the Back of the Yards neighborhood on Chicago's South Side, was started in 2010 by Bubbly Dynamics (BD), a for-profit social enterprise whose mission is "the creation of replicable models for efficiencies that close loops of waste and energy and to encourage others to implement these techniques to combat climate change" [49]. During the first half of the 20th century, Chicago's Back of the Yards district was considered the meatpacking center of the US. The neighborhood grew at the intersection of many railroads and became home to advances in production systems, including the first refrigerated boxcar in 1880, and the country's oldest community organization, Back of the Yards Neighborhood Council (1939). In 1971, the closure of the stockyards was the main trigger for degradation of the neighborhood, including high levels of soil contamination, lack of education, and a high concentration of low-income and marginalized populations.

BD's vision for The Plant was for it to become a model for new sustainable urban food production and consumption systems by transforming an abandoned industrial building into a fully occupied food and beverage business incubator [50]. Over the last decade, The Plant has been home to a collective of urban agriculture and sustainable food start-up or early-stage enterprises, most of them drawn to the facility by its sustainability and circular economy ethos. In 2017–2019, the building housed two dozen tenants including indoor and outdoor farms, kombucha and beer breweries, a cheese distributor, a coffee roaster, and other small food and beverage producers and distributors. Together, they occupied about 80% of the facility and employed 85 full-time employee equivalents. Among the tenants was Plant Chicago (PC), a non-profit formed by BD to lead education and research activities at The Plant, particularly around CE [51]. Among other offerings and programs, PC promoted and operated an on-site farmers market that became a venue for tenants in the building to sell their products and engage with visitors from across the city and around the world, and to a lesser extent, people from the neighborhood.

The renovation of the facility has been gradual. Instead of taking a traditional approach to physical redevelopment, through which high upfront investments determine the quality, allocation, and use of different types of resources, including space design and hard infrastructure within facilities, BD followed a path of adaptive growth. Such an approach resulted in a series of infrastructural interventions that co-evolved in a non-linear manner with the practices they underpin. The company repurposed different parts of the building as new tenants desired to occupy these spaces, or when current tenants requested specific changes based on their operational evolution. This is not only a financial strategy to reduce the uncertainty of resource allocation, as BD uses rents of the finished spaces to reinvest in the building or fund the remodeling of new spaces for incoming tenants, but also an ecological approach informed by principles of complex adaptive systems, such as self-organization and emergence. For example, as new spaces are occupied, additional byproducts become available, giving room for BD to experiment with new technologies (e.g., redirecting carbon dioxide from brewing beer to enhance plant growth in a vertical farm). By having time flexibility and control over the construction process, BD maximizes reuse of the materials in the building.

Overall, the adaptability of The Plant community was shaped by the evolution of interactions between human, technical, and ecological components of the system. As a collaborative, ongoing redevelopment process, the interactions between these elements are key to maintain, augment, or abandon sustainability-oriented practices enacted within the facility. Thus, each infrastructural transformation has built upon and been shaped by previous interventions, including this research and the impacts it has created in the facility.

3.1.2. Focus 1: Quantifying Material Flows, Industrial Symbiosis, and Circular Economy Potential at The Plant

In summer 2016, researchers (W.A., J.P., and other collaborators) conducted the first comprehensive material flow analysis for The Plant, measuring material inputs, electricity use, and waste generation at individual tenant businesses and the facility as a whole [52]. The results of this research highlighted the need for 1) a framework to actively engage small business occupants of the facility in the overall mission of moving towards the circular economy, and 2) an easy-to-use and ongoing system to collect data and measure the sustainability performance of the individual actors and facility as a whole. Follow-up work by BD and tenants focused on building the physical infrastructure for increasing facility-scale industrial symbiosis [50]. For example, inefficient water use was highlighted as a major deficiency in the 2016 study, and BD subsequently designed and installed a rainwater collection system for use by building tenants.

3.1.3. Focus 2: Activating and Mobilizing Agents Advancing Circular Economy at The Plant

PC, BD, and researchers (W.A., A.N., and IIT graduate students) started the current research collaboration to explore sustainability performance measurement frameworks with which the occupants of The Plant could track their material and energy flows. The assumption was that with a framework in place, tenants would engage more in sustainable initiatives because they would have more information regarding the technical and financial viability of engaging in CE practices with The Plant community. Researchers reviewed existing sustainability performance frameworks and reporting tools to determine best practices. In parallel, researchers conducted user and field observations, as well as semi-structured interviews with tenant business owners and staff, BD and PC staff, volunteers, visitors, and several others engaged in The Plant community. These observations revealed key barriers to engaging in and making progress on CE initiatives. They also helped to uncover the priorities of the facility owners, main incentives underlying engagements, and critical challenges concerning individual operations and collective CE efforts. For example, without a formal structure and proper mechanisms to collaborate, current (particularly new) tenants were struggling to understand their role in promoting, leading, or participating in the CE practices proposed by PC and BD.

A series of co-creation workshops were subsequently held with stakeholders within this community to chart a path for deepening and sharing an understanding of CE and the challenges faced by individual companies in enacting it. Collectively, these interactions resulted in the identification of four strategies for achieving this mission, each with its own set of tactics: (1) clarify PC's and BD's vision for CE; (2) engage a broader set of stakeholders; (3) strengthen leadership and support within the facility; (4) measure environmental performance. The staff of BD and PC, as well as some of the tenant business owners within the community, took leadership positions in this endeavor, serving as repositories of technical, business, and tacit knowledge to educate and engage the newer tenants in the facility on the CE journey. A research report was created describing these strategies with tactical recommendations, distributed among tenants and served as the background for the next phase of research [53].

Expanding Perception of the Circular Economy at the Plant

In summer 2017, the researchers (A.N.) built a research base inside The Plant and performed a series of activities that ranged from co-defining CE with tenants, lunch and learn meetings, and design-led workshops with tenants and volunteers, among others. The overall goal was to build new relationships between PC, multiple tenants and their staff, residents of the surrounding neighborhood, volunteers at The Plant, and BD. Researchers used "infrastructuring" to better understand the hard and soft infrastructures conditioning CE practices, as well as the individual activities and business practices that mobilized different types of resources through these infrastructures to actualize CE within the facility. Systems maps were used as physical artifacts to encourage discussion through which different agents in The Plant community could continuously manage concerns and controversies related to implementing CE practices together. In doing so, researchers and participants were able to adapt their activities and consider the unintended consequences of their actions, even as new concerns emerged throughout these engagements.

Interviews with participants alluded to the use of resources beyond the natural, manufactured, and financial resources, such as cultural norms and personal relationships, to actualize CE. To bridge the gap of what is typically considered in CE research, the team explored more expansive frameworks for assessing how different types of resources are utilized by organizations and identifying new leverage points for overcoming the observed barriers to CE adoption [54].

Innovation Lenses Framework

It was in this context that researchers created the innovation lenses framework and developed a series of prototyping activities to explore how the considerations of social, ecological, and technical concerns could be leveraged to increase the effectiveness of CE practices and broaden participation in them. The framework resulted from integrating socio-ecological and socio-technical concerns, considering eight different types of capital, defined as "any type of resource capable of producing additional resources" [55] (p. 165). The capitals can be classed in three broad categories: social, ecological, and technological (see Table 1). The "social" capitals are created through human activities and interactions, these include human, social, political, and cultural capital. Ecological capital is well-defined for only those natural resources that are deemed valuable and monetized, such as fossil fuels, minerals, and water [46]. The "technical" capitals are human-made resources that are only present in economic activities and include financial, digital, and manufactured resources. The framework has been used to describe how different types of resources are used in existing SETS, and to generate interventions that embrace an expanded understanding of the dynamics shaping CE options.

Table 1. Definition of the eight capitals as innovation lenses. Reprint with permission [54]; Copyright 2019, Elsevier.

Social			
Human	**Social**	**Cultural**	**Political**
The ability and capability of individuals to produce and manage their well-being. It includes individual health, knowledge, skills, and motivation.	The professional and social connections among agents. It includes partnerships and collaborations, as well as informal gatherings.	Values and beliefs inherent in social practices, or incorporated by communities, that determine patterns of behavior being encouraged, discouraged, or tolerated by individuals and organizations over time. It also includes ethnicity, spirituality, heritage, traditions, and daily practices.	Governing structures in organizations that determine how decisions are made and power is distributed. It involves hierarchy, inclusion, equity, transparency, access, and participation.
Ecological	**Technical**		
Natural	**Financial**	**Manufactured**	**Digital**
Comprises natural resources, both renewable and nonrenewable. It also includes fauna and flora, as well as their life-supporting systems.	The productive power in the resources of other types of capitals. It includes the resources and assets of an individual or entity translated in the form of a currency that can be accessed, owned, or traded.	All material goods. It includes human-made elements such as physical infrastructures, roads, artifacts, and machines.	Digital infrastructure and data. It includes digital platforms, as well as the mechanisms of data collection, analysis, and storage.

The application of the framework allowed participants to raise questions that reflected a systemic understanding of how different types of resources flow and are allocated based on the individual's activities and business practices within The Plant (see [54]). Participants explored alternative pathways to activate and mobilize available resources and assets for them to contribute to the CE within and beyond the facility's boundaries. For example, PC realized that the value in the human resources in the surrounding neighborhood had not been tapped because of the focus on material loop closing at The Plant, and so sought to bridge practices at the meso-level in the facility, with the needs of surrounding residents. Awareness of the disconnect between achieving "facility-scale Industrial Symbiosis" and investing in residents of the Back of the Yards led tenants at The Plant, including PC and BD, to prioritize hiring and training their neighbors to work inside the building. They also began to support neighborhood entrepreneurial ventures, both by providing infrastructures for them to develop their offerings and by welcoming small neighborhood businesses in their farmers market. Combined, these activities supported PC staff in realizing that their organization was lacking alignment between its goals to incubate local circular economies, current strategies, and its offerings to actualize CE practices. As a response to this challenge, researchers and PC staff collaborated to build each other's capacity to innovate and promote CE practices at the local level and began exploring how to scale impact at higher levels.

3.2. City-level: Farmers Markets

3.2.1. Focus 3: Contextual Background and Problem Framing

Farmers markets provide a platform for farmers to sell their produce directly to consumers. Over time, they have expanded from being a point of sale for farmers to providing novel "farm to fork" food for wealthier consumers and addressing food security and access in low-income communities [56,57]. Most recently, they have also started to engage in "shop local" movements to support small businesses, as well as community building through various complementary programming.

Farmers markets can be characterized as goal-oriented, flexible platforms situated across cities. The POEMS design framework (people, object, environment, messages, and services) defines people (market managers and cleaning staff), objects (wayfinding posters, bins, posters, flyers, tends, etc.), environments (physical space, parking, etc.), messages (information regarding vendors and their produce, incentives to buy local and fresh food, food recipes, etc.), and services (cooking classes, kids tables, prepared food and beverages, live music, etc.) [47,58]. As situated platforms, farmers markets enable the allocation and circulation of various types of resources in urban environments by integrating different production and consumption systems (finance, waste collection, mobility, regulatory, entertainment, food, fashion, health, etc.) that shape and condition various interactions between farmers, local businesses, residents, nonprofits, and government organizations [59]. Yet without proper infrastructures to integrate their efforts into other movements, farmers markets are limited in leading large-scale changes, such as those required to implement CE practices citywide.

Farmers markets hold immense potential to demonstrate, promote, and engage urban businesses and residents with CE practices, as they collectively attract hundreds of thousands of visitors each year in Chicago. Yet there are many barriers, including lack of adequate funding, staffing, information, energy, and waste diversion infrastructure, which make it difficult to implement CE beyond market-level initiatives. Grant funding is one of the chief determinants of what activities and programming market managers can develop and incorporate. Although farmers markets depend upon a large number of interactions between diverse sets of agents, they are isolated within their geographic boundaries. As such, each of these interactions, including among market managers across the city, holds an opportunity for education and action around CE.

3.2.2. Investigating Opportunities for Scaling Niche-Level CE Success to City-Level Impact

The research team (A.N., E.L., and interns from PC) applied the innovation lenses framework to identify how activities performed by PC staff activated and mobilized the eight capitals and to create new strategies for increasing PC's capacity to lead research and innovation activities that scaled the impact of their local CE work.

Given PC's development of a farmers market at The Plant and burgeoning collaborations with other markets throughout the city, the researchers focused on the unrealized value of farmers markets to advance CE in the City of Chicago and explored them as urban infrastructures to expand and replicate local CE practices [60]. Farmers markets have increasingly aimed to support local businesses while also attempting to tackle food insecurity and malnutrition [61]. The challenge in Chicago of how local CE practices should be incorporated into the rules, regulations, and daily operations of the market itself, has yet to be uncovered. Therefore, the focal question for this phase of the research was: What hard and soft infrastructural changes could enable farmers market managers to advance local circular economy practices in Chicago?

3.2.3. Application of the Innovation Lenses Framework in Farmers Markets in Chicago

While farmers markets have been positioned as flexible platforms capable of changing and adapting to address social needs and consumer demands regarding larger societal movements, they operate in isolation. Without proper soft infrastructures to integrate their efforts into other movements, farmers markets are limited in leading large-scale changes, such as those required to implement CE practices citywide. As a result, researchers investigated situations within which farmers markets enabled the integration of multiple systems (e.g., food, mobility, education, health, economy, natural) and supported multiple interactions among diverse sets of agents (e.g., customers, vendors, market managers, hosts, volunteers). To do so, researchers considered each farmers market as a system in itself and conducted ethnographic research on both users and infrastructures through the innovation lenses framework. Table 2 provides a set of questions that guided this phase of research, and examples of the infrastructures, stocks, and flows of different types of resources across them.

PC's staff led most of the activities, such as conducting a literature review on how farmers markets have gone through iterations over time, exploring their contribution, evolution, and adaptation within urban environments during the 19th and 20th centuries. Together, the research team visited different farmers markets, grocery stores, restaurants, urban farms, community centers, and research institutions, and engaged with local vendors, farmers market managers, customers, and peer organizations for participant observations and interviews. At every opportunity, researchers sought to have both formal and informal conversations with diverse agents, including vendors, visitors, farmers, etc., about their personal and professional experiences in farmers markets, and the infrastructure supporting their activities around Chicago.

Upon identifying a variety of resources and assets within farmers markets, the team utilized the anatomy of infrastructures tool (Figure 3) as a structure to organize and make sense of the data gathered. This tool combines principles of multi-level systems mapping, the POEMS design framework, and the innovation lenses framework. It illustrates (1) how different types of resources are flowing through existing offerings; (2) the actionable properties these offerings currently afford users based on the access to specific resources (e.g., affordances); (3) the impacts they generate at different levels considering resource flows and allocation; and (4) their relationships with the main goals.

Table 2. Infrastructures, stocks, and flows of resources in farmers markets across Chicago.

Capital	Dimension	Stocks		Flows	
		Guiding Question	Examples	Guiding Question	Examples
Human	Knowledge	Whose knowledge and labor are considered?	Market managers knowledge of regulations, advertising, and vendors knowledge of produce and goods	How and where is knowledge being created, and for whom?	Market managers analyze data and activities of markets and act accordingly
	Well-being	How is the capacity of individuals to perform defined?	Individuals' health, education	What activities maintain or enhance individuals' capacity to perform?	Cooking classes, healthcare check ups
Social	Professional	Who is considered a partner in farmers markets? What is the nature of the partnership or affiliation?	Suppliers, market vendors, city officials, volunteers	How and where are partnerships being formed?	Outreach of market managers, vendor applications, networking events across the city, vendor training
	Personal	What informal ties exist within current operations?	Relations among managers, vendors, and customers help them to participate in the market	How and where are activities supporting informal gatherings happening?	Informal social gatherings in designated common areas among managers, vendors, and customers
Cultural	Local	What are the local traditions and cultural heritage farmers markets rely upon, and what values and beliefs they sustain?	Local food producers, activists of local and organic movements, language and vocabulary used to communicate with visitors	How and where are the cultural practices and values manifested?	Selection of farmers/vendors/ activities presented at market
	Global	What global elements and practices have been incorporated?	Organic certification/standards, variety of attractions beyond food (e.g., health, music, art, dance), bilingual communications	How and where are new global practices being incorporated?	Market managers host diverse activities, media messaging about food safety and health, local grocery stores pose competition
Political	Regulations	What are the local, state, and federal policies influencing decisions in farmers markets?	City regulations, state/federal food safety rules, funding available for different types of markets, food assistance programs	How and where are policies being enforced or changed?	Market managers have to follow food safety compliance, vendors, and double-value data collection, waste management
	Norms	What is the power structure within current operations?	Market owners/managers make decisions for all participants, vendors, and consumers	How and where are decisions being made, or power shifts taking place?	Decisions are typically made by owners/managers, outside of the market, and presented to vendors prior to the event

Table 2. *Cont.*

Capital	Dimension	Stocks		Flows	
		Guiding Question	Examples	Guiding Question	Examples
Natural	Fauna and Flora	What is the composition of flora and fauna species supporting the farmers markets?	Vegetables and animal products sold, organic and conventionally produced within and outside the city	How are species growth rates affected by the market activities?	Farming practices, organic and conventional, tend to disrupt the ecosystems in which they exist
	Life support systems	What are the energy, materials, and services provided by nature to farmers markets?	Soil for growing produce to be consumed by people or other animals	How and where are energy and nutrients being extracted and regenerated?	Extraction of resources happen within and outside the city, most markets outsource their waste management services
Financial	Services	What institutions provide financial services to agents participating in the markets?	Loans: banks, credit unions; Grants: federal, state, and local governments, foundations; Payments: consumer income, government food assistance	How and where are financial services being provided?	Formal financing of vendor operations; on-site ATMs, SNAP program, double-value services
	Money	What is the institutional structure defining value?	Financial institutions, philanthropic foundations, market competition	How and where are monetary flows occurring?	Outside market: grants, expenses; Inside market: product sale transactions
Manufactured	Infrastructure	What is the physical infrastructure available and its condition?	Transportation networks, automobiles, market owned space furniture, building and city utilities	How and where is physical infrastructure being used and enhanced?	Location of the market determines accessibility and space occupation; seasonal markets may occupy different spaces
	Products and Services	What products, byproducts, and services support activities?	Packaging, marketing materials, educational activities, kids table	How and where are products and services being produced and consumed?	Packaging typically produced elsewhere and discarded at consumers' homes.
Digital	Infrastructure	What are the digital infrastructures available?	Points of sale systems, social media, computers	How and where are digital infrastructure and data used and enhanced?	Data analyzed to gain additional funding
	Data and Information	What data and information supports activities?	Quantified attendance, sales data	How and where are data collected and managed?	Managers and vendors collect data on sales, attendance, promotions

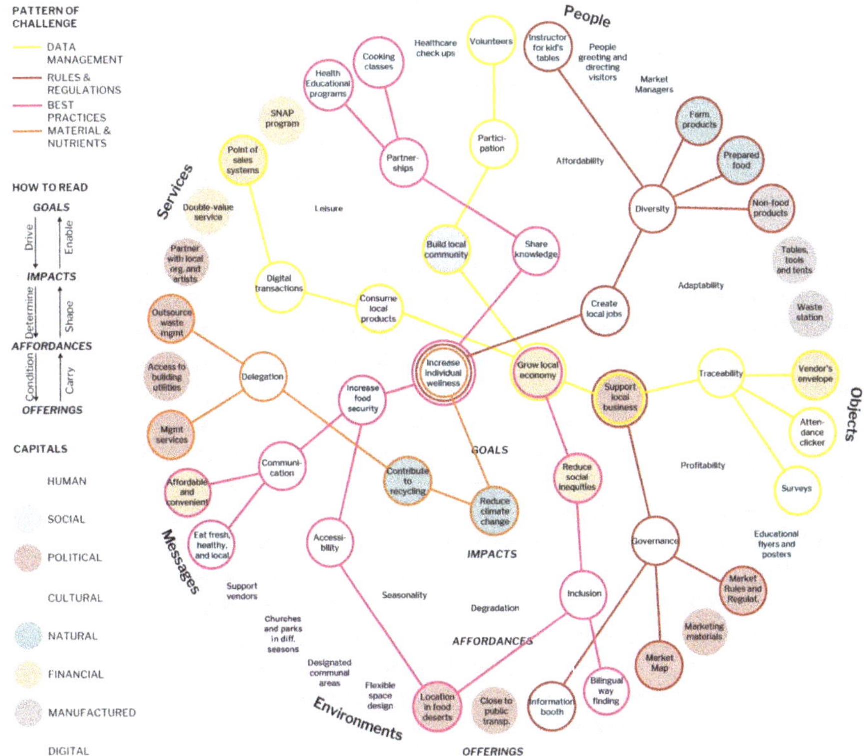

Figure 3. Anatomy of infrastructures underlying farmers markets in the City of Chicago.

The anatomy of infrastructures tool supports the investigation of the overall alignment between the offerings of a system (outer circle) and the intended goals of an infrastructure, situated in the center of the diagram. It considers that resources flow through infrastructures considering actionable properties that their offerings bring to reality, and the impacts users generate given the intended goals. This tool follows the subsequent logic: the goals of man-made infrastructures (center of the diagram) determine the intended and unintended impacts that individuals and organizations create in the broader context in which they exist when they mobilize different resources. These impacts are determined by the possible actions that users can take given the offerings available (outer circles), thereby suggesting how individuals and organizations may use or leverage them through their activities. The correlation between offerings, affordances, impacts, and goals is helpful to understand the current relationships within the hard and soft dimensions of the infrastructure being analyzed, and how these four elements can be integrated into new system interventions.

The team used this tool to understand the infrastructural complexity in advancing CE in farmers markets, and how interventions should consider the integration of infrastructures present in other systems, beyond the farmers market system. For example, the team learned that a set of vendors developed the capacity, relationships, and reputation to sell their products in different markets across the city, but their footprint leaves little room for local or neighborhood businesses to compete within these spaces. Even though diversity is an affordance of current infrastructures that enables a variety of offerings to come to life in each market, the same property is not being leveraged for addressing challenges underlying social inequalities, especially those related to the diversity of vendors across markets.

3.2.4. Patterns of Challenges in Mobilizing Resources

The application of new frameworks and tools led researchers and managers to identify four patterns of challenges related to allocating and mobilizing different types of capital through the farmers market infrastructures.

- Digital and financial capital—managing diverse sets of data: Integrating data from multiple sources and transforming the results into meaningful information requires new competencies within current managerial practices. Farmers markets' managers face the same challenges that managers in large organizations face: what data should be collected? Which sources should be integrated, and to what end? Yet, because farmers markets are usually run by small organizations with resource constraints, they had to develop the minimum necessary set of competencies to create the intended values for specific user groups. For this reason, they often do not have the capabilities to integrate environmental- or social-related data into their operations. This focus revealed an opportunity for collaboration around benchmarking and data sharing to communicate local circular economy practices with each other and with the City government.
- Political and cultural capital—updating rules and regulations: Market managers tend to align their practices and regulations with the larger goals of the City of Chicago programs. Markets must comply with a wide range of rules and regulations, but existing CE practices have emerged organically. Thus, where CE practices exist, they are defined by the vendor's interest and are not explicitly encouraged by the City government. Through this work, the managers are exploring how they can shape new market regulations to encourage the adoption of local circular economy practices by vendors and customers.
- Human and social capital—sharing best practices: Several CE practices have taken root in markets across the city, but opportunities to accelerate and increase impact remain untapped. Chicago's market managers currently share best practices with each other through informal relationships and infrequent gatherings. Without a formal infrastructure to create and sustain more open and inclusive mechanisms to share their learnings and experiences, many managers go through the same struggle at different times. Consequently, the overall farmers market operations become inefficient in resource allocation. This realization led managers to create a systematic way to learn from each other, employ proven successful practices at their markets, and advance the local circular economy.
- Natural and manufactured capital—managing materials and nutrients: Implementing a materials and nutrient management system based on CE practices can significantly increase the diversion of recyclable and biodegradable materials from landfills. In the past five years, a handful of markets have hired small, local food scrap haulers to handle their organic waste, but many markets identified lack of budget, infrastructure, knowledge, and available local services as barriers to diverting materials and nutrients from landfills. Since the majority of waste generated in farmers markets is biodegradable, the ability to implement a system that properly manages materials and nutrients is highly dependent on the market's host site and the priorities of the market's host organization. This insight led to the consideration of how managers could collaborate to create an affordable and effective system to divert materials and nutrients away from landfills.

3.2.5. CE Interventions

The researchers hosted a design-led workshop at the IIT Institute of Design for market managers, representatives of the City of Chicago, and other stakeholders working on CE initiatives from the private sector and non-governmental organizations (NGOs). The goal was to co-create alternative paths to advance local CE through their markets. The workshop represented an action situation (see [18]) necessary to advance CE practices in complex, open-ended projects involving multi-stakeholders, such as urban farmers markets. Each participant received a contextual report created by the researchers indicating the approach to the research and the four common challenges that were uncovered. After validating interpretations about these four challenges, participants divided themselves into small groups, each responsible for one challenge, and considered principles of transparency, diversity, and inclusion as underlying criteria for intervening in current flows and allocation of resources. Participants agreed upon (1) a common goal (advance local circular economy in Chicago through farmers markets), (2) a set of challenges, and (3) the criteria for intervention (principles for local CE) to co-define principles for future engagements and explore new competencies that market managers needed to enable local CE. These priorities and competencies are education (knowledge dissemination), facilitation (CE-oriented interactions to sustain engagement with vendors and customers), collaboration (organize and intentionally support one another), and coordination (ensure that collaboration leads to actions). Combined, these four competencies present opportunity areas for intervention and impact related to building soft infrastructures for the CE at the city-level (see Figure 4).

The understanding and implementation of CE practices at farmers markets require market managers to be connected and communicate regularly. This project served as a starting point for a series of collective efforts to increase the adoption of CE practices currently being led by a coalition of farmers market managers. A research report was co-created and is currently being used by PC staff and other market managers to continue to form new engagements with various stakeholders involved in farmers markets activities [60]. Stakeholder group activities currently include advising the City of Chicago on market regulations, changing the rules and regulations of individual farmers markets programs, maintaining a digital platform for market managers to communicate, and building vendor's capacity to engage in CE practices, among others. This highlights the opportunity for scaling the impact of meso-level successes, by pivoting through strategic alliances, systems understanding, and infrastructuring co-design practices with lead agents and partners.

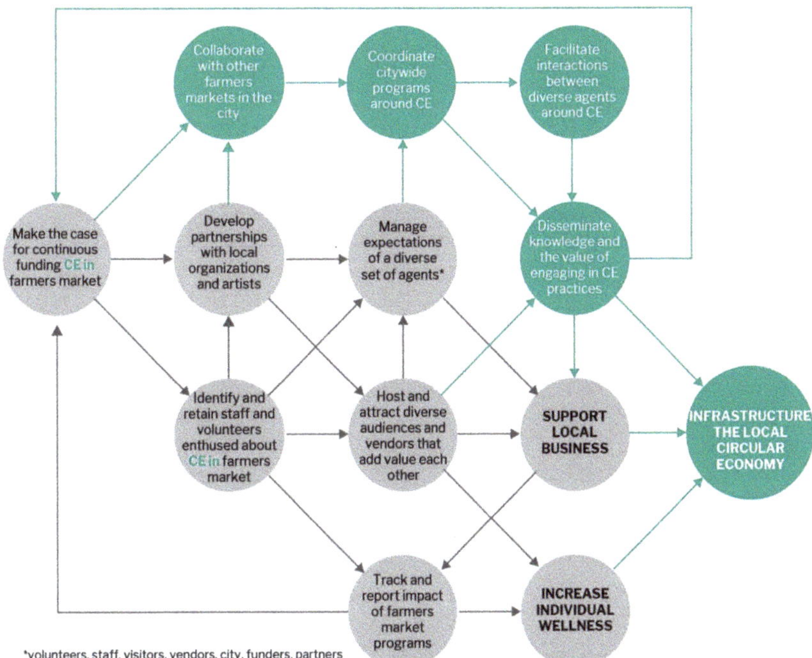

Note: Grey elements illustrate existing competencies needed to bring a farmers market to life, and their interdependencies (arrows). Green elements (arrows, texts, and circles) situate the changes and the four new competencies identified by researchers and participants.

Figure 4. Competencies needed for infrastructuring the local CE through farmers markets.

3.2.6. Four I's Model

Combined, the experiences at the facility-level (The Plant) and the city-level (farmers markets in Chicago) culminated in the "Four I's" model, which presents four intervention strategies needed in any process of "infrastructuring" the CE practices in SETS, regardless of its geographic boundary. The "Four I's" model consists of: (1) interconnectivity between the organizational levels of systems; (2) integration of the social, ecological, and technical systems shaping the conditions in place; (3) interactions of the agents defining the dynamics of these systems; and (4) iteration of design interventions over time [62]. The Four I's model ultimately presents four general design strategies for intervening in complex CE challenges, such as in urban environments. For each one of these strategies, researchers are exploring a specific tool that can support and advance design practices in these complex spaces.

Interconnectivity of Organizational Levels

People experience different conditions of SETS and develop unique knowledge about them depending on the level within which they are embedded (macro, meso, micro). Unlike embedded (operational) and explicit (codified) knowledge, "tacit knowledge can neither be explained in terms of rational decision-making nor be summarized easily in quantitative terms" [63] (p. 55). This challenges traditional approaches to expertise and requires different stakeholders to understand the value of daily life experiences that manifests itself as tacit knowledge, to inform change. Thus, in addition to capturing the different types of knowledge about the conditions in each one of these levels, individuals and organizations need new approaches that recognize and incorporate considerations of how the interdependency among them is shaping the dynamics of SETS. Although activities at each level need certain autonomy to increase efficiency and effectiveness, they also need to be connected and integrated with the choices and activities happening at the other levels. Without proper alignment, the chances of unintended consequences might increase because a choice made in one level invariably will be made based on the unrealistic assumption that the other levels will support or are capable of adapting accordingly.

Integration of Multiple Systems

A system is a collection or set of interconnected parts, usually delimited by some type of spatial or temporal boundaries. Systems' boundaries can be outlined based on the structure, the functionality, the nature, and/or the intended goal of the analysis and design [64]. As individuals and organizations focused on creating new CE practices in cities tend to interpret systems' boundaries of their innovation processes based on the specialized practices of their own industry or sector, they create artificial boundaries for interventions that are defined by their focal idea, not by the natural or existing limits of the system. Yet, infrastructural interventions in urban areas result from individual activities and organizational practices that activate and mobilize various types of resources currently distributed across multiple systems (e.g., energy, knowledge, water, money, power, etc.). Thus, when these agents take traditional approaches that consider one system at a time, they overlook the extent through which infrastructural interventions are conditioned by, and dependent on, intersecting systems that shape everyday life [65].

Interactions of Diverse Sets of Agents

The interaction of diverse agents can bring distinctive perspectives to framing problems and developing solutions to CE challenges in cities. Agents can be human or non-human, including components of technology (e.g., portable machines, digital platforms, organizations, products, etc.) or biological (living creatures, etc.). An interaction in SETS occurs when any agent affects the other. Interactions can be defined as symbiotic relationships between and among social, technical, and ecological components. They are dynamic relations between parts and wholes in systems and can be transactional (e.g., a single purchase from a vendor at a farmers market) or defined by a temporal pattern (e.g., increase of visitors at The Plant during wintertime). Redesigning the flows of political capital within various systems that shape local dynamics should underlie efforts to create infrastructural interventions. Currently, this resource is unevenly distributed and hoarded by certain agents within SETS. Such conditions create unsustainable power dynamics among agents and influence the outcomes of infrastructural interventions. Boonstra defines power as "a (human) capacity to act in social and ecological conditions" [66] (p. 1). Thus, understanding the role of power in shaping contemporary socio-technical and socio-ecological interactions in urban territories is critical to map and intervene in current urban dynamics. As noted by Geels, existing institutional arrangements operate within certain power dynamics, and to intervene in them, individuals and organizations that seek to create urban CE practices need to understand who has the power to enable or inhibit large-scale transformations, including how the relationships among and between these two groups are sustained [67].

Iterations over Time

Having considered the dynamic interactions of diverse sets of agents as the third attribute of infrastructural intervention, this last part addresses the iterative nature of CE interventions. Infrastructural interventions "are always and unavoidably situated within, and part of the sedimentation of material arrangements, themselves linked to a persistently dynamic profile of activities and practices" [65] (p. 164). Infrastructuring the CE demands recognition that infrastructures are path-dependent and impermanent because of the variation of the activities they enable. The dynamic nature of everyday individual activities and organizational practices suggests that intervening in existing infrastructures to support local CE demands agents involved in these processes to work in successive intervention processes, as the interactions between diverse sets of agents utilizing these infrastructures will necessarily change over time.

Cities, as SETS, are shaped by interdependent infrastructures that produce synergies due to complementary functions they perform in activating and mobilizing different types of resources. Such interdependence will likely support the co-evolution of both problems and solutions to city dwellers, organizations, and ecosystems within the urban territory. Thus, infrastructural interventions are dependent on how different resources distributed across social, technical, and ecological systems can be activated and mobilized so as to serve one or more SET demands (e.g., Bubbly Dynamics repurposing a building, while contributing to local economic development). This work suggests that diverse agents driving CE change in cities need to adopt more pliant approaches that can enable them to continuously engage and involve diverse voices representatives of the plurality of residents and local business in both processes of problem framing and solution finding, as well as in the resulting pathways that lead to creating and implementing infrastructural interventions for local CE practices.

4. Discussion

A significant gap remains to connect conceptual and ideological discourses of CE practices to the pragmatism required to intervene in complex situations that need reconsideration of the allocation and circulation of different types of resources, so they are fit to the sustainability of the SETS of interest. Transitions in SETS, such as cities, require knowledge integration from both multiple disciplines and diverse agents distributed across the different levels of these systems. As knowledge is dispersed in multiple forms and agents shape the dynamics in SETS, new approaches are required to not only increase impact in each one of these systems but also to carve new opportunities and explore emerging possibilities. This work focused on creating innovative infrastructural interventions centered on overcoming complex socio-ecological-technical challenges. More specifically, it focused on expanding CE practices to contribute to the sustainability of the systems within which they exist and advancing expertise in providing alternative, regenerative approaches to scale impact from niche-level successes to higher levels in SETS.

In the case of both the facility- and city-level, the opportunity relied on the creation of new knowledge infrastructures, or "robust networks of people, artifacts, and institutions that generate, share, and maintain specific knowledge about the human and natural worlds [68] (p. 5). Here, knowledge infrastructures are recognized as adaptive and involving a continuous flow of information because (1) individual elements are constantly changing, leaving, or even being introduced, and (2) "knowledge is perpetually in motion" [64] (p. 6), meaning that the definition of the known is constantly changing either by novel questions, redefinitions, or incorporations of novel perspectives, which was the case of the CE paradigm. Such fluid conditions frame knowledge as a resource flowing through the interactions between different agents in SETS. As Edwards and colleagues argued, "the current situation for knowledge infrastructures is characterized by rapid change in existing systems and introduction of new ones, resulting in severe strains on those elements with the greatest inertia" [68] (p. 5). Thus, if individuals and organizations interested in enabling CE practices intend to capture these resources and embody them into new interventions, they will have to develop new mechanisms to integrate knowledge that is not only pulled from different domains (e.g., industrial ecology, socio-technical, design, and socio-ecological systems theory) but also distributed among different agents (residents, local institutions, researchers, local businesses, investors, government agencies, etc.), considering how they are interacting on multiple levels of the SETS of interest. By doing so, individuals and organizations may be better equipped to design alternative infrastructures capable of confronting complex socio-ecological-technical challenges preventing the wide adoption of CE practices within urban environments.

As urban change makers, researchers working with individuals and organizations embedded in cities leveraged the innovation lenses framework and the Four I's model to explore infrastructure within The Plant and in farmers markets across Chicago. Both were approached as junctions of social, ecological, and technical systems enabling the allocation and movement of multiple types of resources. Upon completing the first comprehensive material flow analysis of The Plant, our research focused on co-creating a sustainability measurement framework to support The Plant's tenants in their engagement with CE activities. By applying prototyping methods as means for infrastructuring change through PAR, the team uncovered the need, and opportunity, to broaden participation and involvement in determining and shaping CE activities both within and beyond the facility (the niche), and to consider the variety of perspectives (Four I's model) and resources (innovation lenses framework) that shape sustainability and equity within SETS dynamics. As a response, Plant Chicago repositioned itself as a convener for local circular economy practices, facilitating niche-to-regime transitions towards circularity across the city, in this case, through farmers markets.

This research project not only helped build design capacity among PC staff, increasing their ability to embed principles of sustainability and equity into their innovation processes towards CE, but it also increased their confidence in tackling complex projects at a broader scale, beyond the facility boundaries. In the process of doing so, PC transitioned from an NGO that promoted CE through research and development at The Plant to one cultivating local circular economies across the City of Chicago by convening diverse perspectives that would not otherwise have a forum in the circular economy space. Researchers, on the other hand, benefited from social connections and political access to various communities, and from the knowledge and experience of diverse agents embedded in these communities as they had deep expertise about existing infrastructural challenges, as well as historical and cultural patterns that continue to prevent CE to be actualized. Such contributions were keen to advance on new approaches, combining frameworks, methods, models, and tools that can facilitate and contribute to the sustainability and equity of SETS dynamics through CE practices. Other agents directly or indirectly involved in this research benefited from the activism and leadership of PC to intervene in the flows of resources through infrastructural interventions.

The researchers, PC staff, and market managers explored farmers markets as citywide infrastructures to promote and enable local CE practices, rather than isolated platforms. Together with other market managers, the research team realized that by incentivizing changes in the business practices of small companies involved in farmers markets, vendors will likely change their practices elsewhere, consequently creating a network effect across the city with the farmers market as a leverage point for change. As urban infrastructures are (re)oriented towards CE, farmers markets in Chicago present high potential to become the means through which other agents learn about and can practice CE. Currently, they are not only becoming regulated by CE principles, but also provide incentives for the adoption of CE practices through various activities (e.g., capacity building of vendors, periodic meetings of market managers), aligned messages (e.g., posters informing opportunities and decisions around CE, more inclusive and diverse messaging to reflect the population it serves, etc.), and new offerings (e.g., new waste stations, compost services, advice and recommendations for policy change in the City of Chicago, among others). By relying on the concept of infrastructuring, market managers are creating new CE-focused programs that increase interactions between small local businesses, while also forming a local network centered on CE. In doing so, they are connecting data, information, and strategies from attempts in each market, sharing with each other, and co-creating and prototyping alternative interventions based on the resources available and on the opportunities to promote well-coordinated, citywide change towards CE.

5. Conclusions

The application of the innovation lenses framework and Four I's model can help agents in public, private, and civil society to consider how infrastructural interventions activate, mobilize, and are conditioned by the flow and allocation of various types of resources, the interconnectivity of different organizational levels, the intersection of social, ecological, and technical systems, the interactions of diverse sets of agents, and their iterations over time. Urban infrastructures are the means through which resources flow in a system, given a specific goal, traditionally to support economic growth through the lenses of technical systems. Today, there is increasing recognition that they have to be adapted and used to support more sustainable and equitable outcomes in urban environments; but to do so, urban change makers must develop their expertise in creating the infrastructural interventions affecting the circulation of different types of resources, and contribute to the fitness of humans and non-human agents' interactions happening within and across systems levels.

Exploring infrastructure from a multi-level, relational perspective unlocks new opportunities for situated urban interventions to consider how the relationships between different agents embedded in these situations are the means through which different types of resources are activated and mobilized within a city. Expanding the traditional boundaries of infrastructural problems to include dynamic interactions of diverse agents underpinning resources flow also enables urban change makers to better understand how these dynamics in turn shape and condition situated interventions. When explored as elements of both socio-ecological and socio-technical systems, infrastructural interventions for CE in urban settings become the nexus through which different types of resources are combined to generate transformational change towards sustainability and equity.

Author Contributions: Conceptualization, A.N., W.A. and C.T.; methodology, A.N., W.A., C.T., E.L. and J.P.; formal analysis, A.N.; investigation, A.N. and E.L.; resources, A.N., W.A., C.T., E.L. and J.P.; data curation, A.N.; writing—original draft preparation, A.N. and W.A.; writing—review and editing, A.N., W.A., C.T., E.L. and J.P.; visualization, A.N.; supervision, W.A. and C.T.; project administration, A.N., W.A. and C.T.; funding acquisition, W.A., C.T. and A.N. All authors have read and agreed to the published version of the manuscript.

Funding: This research was funded by Illinois Institute of Technology's Education and Research Initiative Fund, grant number 17-0344 and the APC was funded by MDPI.

Acknowledgments: The authors wish to thank all the stakeholders at The Plant, in particular John Edel, Carolee Kokola, and Alex Enarson of Bubbly Dynamics, Rosanna Lloyd and the employees of Just Ice, Art Jackson and the employees of Pleasant House Bakery, Adam Pollack and the employees of Closed Loop Farms, Alex Poltorak and the employees of Urban Canopy, Kaley Donewald from Secred Serve, Brian Taylor and Charla from Whiner Brewery, Ria Neri from Whiner Brewery and Four Letter Word Coffee, Matt Lancor from Kombutchade, for generously sharing their time and knowledge. We also wish to thank the Spring 2017 Sustainability Management students: Roshni Lad, John Lerczak, Megan Mckitterick, Flavio Santos Urteaga, and Tian Wen, and Plant Chicago Summer 2017 interns: Tommy Straus and Tyler Washington, who worked with us to interview stakeholders, collect and organize data.

Conflicts of Interest: E.L. and J.P. are employed by Plant Chicago, the non-government organization featured in the manuscript. A.N. and W.A. are members of Plant Chicago's Circular Economy Advisory Board.

References

1. World Economic Forum; Ellen MacArthur Foundation; McKinsey & Company. *Towards the Circular Economy: Accelerating the Scale-Up across Global Supply Chains*; World Economic Forum: Geneva, Switzerland, 2014.
2. Bocken, N.M.P.; De Pauw, I.; Bakker, C.; Van der Grinten, B. Product design and business model strategies for a circular economy. *J. Ind. Prod. Eng.* **2016**, *33*, 308–320. [CrossRef]
3. Ellen MacArthur Foundation Circular Cities. Available online: https://www.ellenmacarthurfoundation.org/our-work/activities/circular-economy-in-cities (accessed on 12 February 2020).
4. Agudelo-Vera, C.M.; Mels, A.R.; Keesman, K.J.; Rijnaarts, H.H.M. Resource management as a key factor for sustainable urban planning. *J. Environ. Manag.* **2011**, *92*, 2295–2303. [CrossRef] [PubMed]
5. Markolf, S.A.; Chester, M.V.; Eisenberg, D.A.; Iwaniec, D.M.; Davidson, C.I.; Zimmerman, R.; Miller, T.R.; Ruddell, B.L.; Chang, H. Interdependent infrastructure as linked social, ecological, and technological systems (SETSs) to address lock-in and enhance resilience. *Earths Future* **2018**, *6*, 1638–1659. [CrossRef]
6. Duranton, G.; Puga, D. Micro-foundations of urban agglomeration economies. In *Handbook of Regional and Urban Economics*; Working Paper 9931; National Bureau of Economic Research: Cambridge, MA, USA, 2003.
7. Sennett, R. *Building and Dwelling: Ethics for the City*, 1st American ed.; Farrar, Straus and Giroux: New York, NY, USA, 2018; ISBN 978-0-374-20033-6.
8. Carpenter, S.R.; DeFries, R.; Dietz, T.; Mooney, H.A.; Polasky, S.; Reid, W.V.; Scholes, R.J. Millennium ecosystem assessment: Research needs. *Science* **2006**, *314*, 257–258. [CrossRef]
9. Millennium Ecosystem Assessment. Available online: https://www.millenniumassessment.org/en/Global.html (accessed on 12 February 2020).
10. Moreau, V.; Sahakian, M.; Van Griethuysen, P.; Vuille, F. Coming full circle: Why social and institutional dimensions matter for the circular economy. *J. Ind. Ecol.* **2017**, *21*, 497–506. [CrossRef]
11. Block, D.; Rosing, H. *Chicago: A Food Biography*; Rowman & Littlefield: Lanham, MD, USA, 2015; ISBN 978-1-4422-2726-2.
12. Block, D.R.; Chavez, N.; Allen, E.; Ramirez, D. Food sovereignty, urban food access, and food activism: Contemplating the connections through examples from Chicago. *Agric. Hum. Values* **2011**, *29*, 203–215. [CrossRef]
13. Caron, P.; Ferrero y de Loma-Osorio, G.; Nabarro, D.; Hainzelin, E.; Guillou, M.; Andersen, I.; Arnold, T.; Astralaga, M.; Beukeboom, M.; Bickersteth, S.; et al. Food systems for sustainable development: Proposals for a profound four-part transformation. *Agron. Sustain. Dev.* **2018**, *38*, 41. [CrossRef]
14. Zumkehr, A.; Campbell, J.E. The potential for local croplands to meet US food demand. *Front. Ecol. Environ.* **2015**, *13*, 244–248. [CrossRef]
15. Heller, M.C.; Keoleian, G.A. Greenhouse gas emission estimates of U.S. dietary choices and food loss. *J. Ind. Ecol.* **2015**, *19*, 391–401. [CrossRef]
16. Bozeman, J.F.; Ashton, W.S.; Theis, T.L. Distinguishing environmental impacts of household food-spending patterns among U.S. demographic groups. *Environ. Eng. Sci.* **2019**, *36*, 763–777. [CrossRef]
17. Clark, M.A.; Springmann, M.; Hill, J.; Tilman, D. Multiple health and environmental impacts of foods. *Proc. Natl. Acad. Sci. USA* **2019**, *116*, 23357–23362. [CrossRef] [PubMed]
18. Ostrom, E. A General framework for analyzing sustainability of social-ecological systems. *Science* **2009**, *325*, 419–422. [CrossRef]
19. Malecki, E.J. Hard and soft networks for urban competitiveness. *Urban Stud.* **2002**, *39*, 929–945. [CrossRef]
20. Sovacool, B. Hard and soft paths for climate change adaptation. *Clim. Policy* **2011**, *11*, 1177–1183. [CrossRef]
21. Larkin, B. The politics and poetics of infrastructure. *Annu. Rev. Anthropol.* **2013**, *42*, 327–343. [CrossRef]
22. Edwards, P.N. Infrastructure and modernity: Force, time, and social organization in the history of sociotechnical systems. *Mod. Technol.* **2003**, *1*, 185–226.
23. Star, S.L. The sociology of the invisible: The primacy of work in the writings of Anselm Strauss. In *Social Organization and Social Process: Essays in Honor of Anselm Strauss*; Transaction Publishers: Piscataway, NJ, USA, 1991; pp. 265–283.

24. Jewett, T.; Kling, R. The dynamics of computerization in a social science research team: A case study of infrastructure, strategies, and skills. *Soc. Sci. Comput. Rev.* **1991**, *9*, 246–275. [CrossRef]
25. Latour, B.; Hermant, E.; Shannon, S. *Paris: Invisible City*; Institut Synthélabo pour le progrès de la connaissance: Paris, France, 1998.
26. Star, S.L. The ethnography of infrastructure. *Am. Behav. Sci.* **1999**, *43*, 377–391. [CrossRef]
27. Star, S.L.; Ruhleder, K. Steps toward an Ecology of Infrastructure: Design and Access for Large Information Spaces. *Inf. Syst. Res.* **1996**, *7*, 111–134. [CrossRef]
28. Bonato, D.; Orsini, R. Urban circular economy: The new frontier for european cities' sustainable development. In *Sustainable Cities and Communities Design Handbook*; Butterworth-Heinemann: Oxford, UK, 2018; pp. 235–245.
29. Petit-Boix, A.; Leipold, S. Circular economy in cities: Reviewing how environmental research aligns with local practices. *J. Clean. Prod.* **2018**, *195*, 1270–1281. [CrossRef]
30. La Recyclerie. Available online: http://www.larecyclerie.com/ (accessed on 12 February 2020).
31. Recycle Here! Detroit. Available online: http://www.recyclehere.net/ (accessed on 13 February 2020).
32. Amsterdam Policy: Circular Economy. Available online: https://www.amsterdam.nl/en/policy/sustainability/circular-economy/ (accessed on 12 February 2020).
33. Circular Charlotte. *Envision Charlotte*; Circular Charlotte: Charlotte, NC, USA, 2018.
34. Wang, N.; Lee, J.C.K.; Zhang, J.; Chen, H.; Li, H. Evaluation of Urban circular economy development: An empirical research of 40 cities in China. *J. Clean. Prod.* **2018**, *180*, 876–887. [CrossRef]
35. Ezzat, A.M. Sustainable development of seaport cities through circular economy: A comparative study with implications to Suez Canal corridor project. *Eur. J. Sustain. Dev.* **2016**, *5*, 509. [CrossRef]
36. Ozanne, J.L.; Saatcioglu, B. Participatory Action Research. *J. Consum. Res.* **2008**, *35*, 423–439. [CrossRef]
37. Klerkx, L.W.A.; Van Mierlo, B.; Leeuwis, C. Evolution of systems approaches to agricultural innovation: Concepts, analysis and interventions. In *Farming Systems Research into the 21st Century: The New Dynamic*; Springer: Dordrecht, The Netherlands, 2012; pp. 457–483. ISBN 978-94-007-4502-5.
38. Forlano, L. Decentering the Human in the Design of Collaborative Cities. *Des. Issues* **2016**, *32*, 42–54. [CrossRef]
39. Ehn, P. Participation in design things. In Proceedings of the Participatory Design Conference 2008, Bloomington, IN, USA, 30 September–4 October 2008; ACM: New York, NY, USA.
40. Baum, F.; MacDougall, C.; Smith, D. Participatory action research. *J. Epidemiol. Community Health* **2006**, *60*, 854–857. [CrossRef]
41. Chevalier, J.M.; Buckles, D. *Participatory Action Research: Theory and Methods for Engaged Inquiry*; Routledge: Abingdon, UK, 2013; ISBN 978-0-415-54031-5.
42. Teram, E.; Schachter, C.L.; Stalker, C.A. The case for integrating grounded theory and participatory action research: Empowering clients to inform professional practice. *Qual. Health Res.* **2005**, *15*, 1129–1140. [CrossRef] [PubMed]
43. Nogueira, A.; Teixeira, C.; Ashton, W.S. *Innovation Lenses Playbook*; Illinois Institute of Technology: Chicago, IL, USA, 2018.
44. Ashton, W.; Lifset, R. Industrial Ecology. In *Encyclopedia of Sustainability, Vol. 4: Natural Resources and Sustainability*; Vasey, D., Christensen, K., Shen, L., Thompson, S., Eds.; Berkshire: Great Barrington, MA, USA, 2012.
45. Ashton, W. Understanding the Organization of Industrial Ecosystems: A Social Network Approach. *J. Ind. Ecol.* **2008**, *12*, 34–51. [CrossRef]
46. Meadows, D. *Thinking in Systems*; Chelsea Green Publishing: Windsor, VT, USA, 2008.
47. Whitney, P. Design and the Economy of Choice. *She Ji J. Des. Econ. Innov.* **2015**, *1*, 58–80. [CrossRef]
48. Bertola, P.; Teixeira, J.C. Design as a knowledge agent: How design as a knowledge process is embedded into organizations to foster innovation. *Des. Stud.* **2003**, *24*, 181–194. [CrossRef]
49. Bubbly Dynamics LLC. Available online: https://www.bubblydynamics.com (accessed on 13 February 2020).
50. Mulrow, J.S.; Derrible, S.; Ashton, W.S.; Chopra, S.S. Industrial Symbiosis at the Facility Scale. *J. Ind. Ecol.* **2017**, *21*, 559–571. [CrossRef]
51. Plant Chicago. Available online: https://plantchicago.org/ (accessed on 13 February 2020).
52. Chance, E.; Ashton, W.; Pereira, J.; Mulrow, J.S.; Norberto, J.; Derrible, S.; Guilbert, S. The Plant—An experiment in urban food sustainability. *Environ. Prog. Sustain. Energy* **2018**, *37*, 82–90. [CrossRef]

53. Ashton, W.; Nogueira, A.; Lad, R.; Lerczak, J.; McKitterick, M.; Urteaga, F.; Wen, T. *Implementing the Circular Economy at The Plant*; Illinois Institute of Technology & Plant Chicago: Chicago, IL, USA, 2017.
54. Nogueira, A.; Ashton, W.S.; Teixeira, C. Expanding perceptions of the circular economy through design: Eight capitals as innovation lenses. *Resour. Conserv. Recycl.* **2019**, *149*, 566–576. [CrossRef]
55. Flora, C.; Flora, J.; Fey, S. *Rural Communities: Legacy and Change*, 2nd ed.; Westview Press: Boulder, CO, USA, 2004.
56. Siegner, A.; Sowerwine, J.; Acey, C. Does urban agriculture improve food security? Examining the nexus of food access and distribution of urban produced foods in the United States: A systematic review. *Sustainability* **2018**, *10*, 2988. [CrossRef]
57. Morales, A.; Kettles, G. Healthy food outside: Farmers' markets, taco trucks, and sidewalk fruit vendors. *J. Contemp. Health Law Policy* **2009**, *26*, 20.
58. Kumar, V. *101 Design Methods: A Structured Approach for Driving Innovation in Your Organization*; John Wiley & Sons: Hoboken, NJ, USA, 2012.
59. George, D.R.; Kraschnewski, J.L.; Rovniak, L.S. Public health potential of farmers' markets on medical center campuses: A case study from Penn State Milton S. Hershey Medical Center. *Am. J. Public Health* **2011**, *101*, 2226–2232. [CrossRef] [PubMed]
60. Nogueira, A.; Lyon, E. *The Future of Farmers Market*; Plant Chicago: Chicago, IL, USA, 2018.
61. Jones, P.; Bhatia, R. Supporting equitable food systems through food assistance at farmers' markets. *Am. J. Public Health* **2011**, *101*, 781–783. [CrossRef] [PubMed]
62. Nogueira, A. Sustainable Solutions in Complex Spaces of Innovation. Ph.D. Thesis, Illinois Institute of Technology, Chicago, IL, USA, 2019.
63. Heskett, J. Design from the Standpoint of Economics/Economics from the Standpoint of Design. In *A John Heskett Reader: Design, History, Economics*; Bloomsbury: London, UK, 2016.
64. Holland, J.H. *Signals and Boundaries: Building Blocks for Complex Adaptive Systems*; MIT Press: Cambridge, MA, USA, 2012.
65. Cass, N.; Schwanen, T.; Shove, E. Infrastructures, intersections and societal transformations. *Technol. Forecast. Soc. Change* **2018**, *137*, 160–167. [CrossRef]
66. Boonstra, W. Conceptualizing power to study social-ecological interactions. *Ecol. Soc.* **2016**, *21*, 21. [CrossRef]
67. Geels, F. Regime resistance against low-carbon transitions: Introducing politics and power into the multi-level perspective. *Theory Cult. Soc.* **2014**, *31*, 21–40. [CrossRef]
68. Edwards, P.N.; Jackson, S.J.; Chalmers, M.K.; Bowker, G.C.; Borgman, C.L.; Ribes, D.; Burton, M.; Calvert, S. *Knowledge Infrastructures: Intellectual Frameworks and Research Challenges*; University of Michigan: Ann Arbor, MI, USA, 2013.

Article

A Hybrid Methodology to Study Stakeholder Cooperation in Circular Economy Waste Management of Cities

P. Giovani Palafox-Alcantar *, Dexter V. L. Hunt and Chris D. F. Rogers

Department of Civil Engineering, School of Engineering, University of Birmingham, Birmingham B15 2TT, UK; d.hunt@bham.ac.uk (D.V.L.H.); c.d.f.rogers@bham.ac.uk (C.D.F.R.)

* Correspondence: p.g.palafoxalcantar@bham.ac.uk

Received: 20 February 2020; Accepted: 6 April 2020; Published: 10 April 2020

Abstract: Successful transitioning to a circular economy city requires a holistic and inclusive approach that involves bringing together diverse actors and disciplines who may not have shared aims and objectives. It is desirable that stakeholders work together to create jointly-held perceptions of value, and yet cooperation in such an environment is likely to prove difficult in practice. The contribution of this paper is to show how collaboration can be engendered, or discord made transparent, in resource decision-making using a hybrid Game Theory approach that combines its inherent strengths with those of scenario analysis and multi-criteria decision analysis. Such a methodology consists of six steps: (1) define stakeholders and objectives; (2) construct future scenarios for Municipal Solid Waste Management; (3) survey stakeholders to rank the evaluation indicators; (4) determine the weights for the scenarios criteria; (5) reveal the preference order of the scenarios; and (6) analyse the preferences to reveal the cooperation and competitive opportunities. To demonstrate the workability of the method, a case study is presented: The Tyseley Energy Park, a major Energy-from-Waste facility that treats over two-thirds of the Municipal Solid Waste of Birmingham in the UK. The first phase of its decision-making involved working with the five most influential actors, resulting in recommendations on how to reach the most preferred and jointly chosen sustainable scenario for the site. The paper suggests a supporting decision-making tool so that cooperation is embedded in circular economy adoption and decisions are made optimally (as a collective) and are acceptable to all the stakeholders, although limited by bounded rationality.

Keywords: circular economy; cooperation; game theory; multi-criteria decision analysis; scenario analysis

---

1. Introduction

A growing body of research suggests that a Circular Economy (CE) approach results in more efficient use of materials and better waste management processes in which resources are continually fed back into the consumption process, rather than reaching end-of-life. CE principles involve resources and waste being reintroduced into the process (indefinitely) rather than effectively becoming lost [1]. As such, it is considered as the opposite of the current linear consumption system. Adoption of a CE approach is fraught with a plethora of associated barriers that need to be overcome [2], not least the ability to:

- Capture multiple value perceptions [3];
- Facilitate cooperation between stakeholders [4];
- Ensure stakeholders are proactive and cooperative in terms of considering and adopting new supply chains [5];

- Raise awareness and provide regulations that support CE [6];
- Achieve an outcome that satisfies (i.e., is welcomed or tolerated by) all participants.

In essence, it is about facilitating a decision-making process that acts as a CE enabler by overcoming the barriers previously highlighted. Multiple methodologies have been reviewed that facilitate decision-making processes for topics such as Municipal Solid Waste (MSW), bioenergy and Industrial Symbiosis (IS)—all relevant aspects of waste management (e.g., [2,7–9]).

The method presented in this paper embraces the advantages of three diverse methodologies into a hybrid approach, namely:

1. Scenario Analysis (SA);
2. Multi-Criteria Decision Analysis (MCDA); and
3. Game Theory (GT).

SA has been commonly used to make future predictions, for example: to identify optimisation measures in the waste household appliance recovery industry [10]; to predict the total greenhouse gas emissions of multiple MSW scenarios [11]; to analyse the influencing factors in MSW scenarios, thereby improving opportunities and identifying key problems [12]; and to compare the economic and environmental impact, and review the energy efficiency, of traditional technologies with mechanical-biological MSW treatment [13]. Additionally, SA and MCDA were used complementarily to find the best solutions of MSW strategies for future scenarios [14].

MCDA techniques have been widely applied to CE studies, a few relevant examples being: a weighting method was introduced to involve stakeholders in the selection process of a MSW facility [15]; MSW studies which focused on the perceptions of stakeholders have been reviewed [16]; subjective preferences of stakeholders and objective performance of eco-industrial thermal power plants were integrated to determine criteria rankings [17]; the disassembly of aircraft at their end-of-life were studied as a MCDA issue [18]; the preferences of alternatives to new uses for waste in mining sites were assessed [19]; and alternatives to import liquefied natural gas whilst satisfying CE-related logistics criteria were optimised [20].

On the other hand, GT elements are less commonly applied, although examples of CE and solid waste studies do exist. For example, the trade-offs between disposable and refillable bottles were studied [21], in which consumers and bottles were incorporated as the stakeholders; the characteristics of Cost-Benefit Analysis (CBA), Life-Cycle Assessment (LCA) and MCDA were contrasted to further introduce a decision support framework based on GT [22]; the optimal alternative from multiple waste-to-energy solutions was selected [9]; and cooperative costs and legislation constraints were included in a study of MSW separation mechanisms [23].

MCDA is used to model the preferences of stakeholder groups in decision-making by introducing 'compensation', meaning to agree on a set of trade-offs which settle for fewer features of the most preferable scenario and more of the less preferable ones, without decreasing the general satisfaction of stakeholders [16]. Whilst GT is able to analyse trade-offs by considering potential cooperation and conflict between stakeholders, MCDA techniques do not consider stakeholders' preferences and their influence when negotiating and attempting to reach consensus [9]. This is a shortfall and, therefore, the potential of combining SA with MCDA and GT offers significant advantages, particularly in the case of the CE. For a more detailed discussion of these, refer to [2].

Additionally, GT elements were used to study group decision-making for landfill and Energy-from-Waste (EfW) technology alternatives [9], whereas this paper has included other CE principles such as reducing MSW generation, recycling and carbon emissions mitigation. In addition, a two-player game was introduced in this study, whilst the framework reported herein considers an expanded n-player game where five stakeholders are considered for the case study provided. The proposed methodology aims to deliver recommendations on how to reach a 'most optimal' scenario. (That is, each stakeholder might have an 'optimal scenario', but for the stakeholders as a whole, there will be a 'combined optimal', which will (at least for some) be 'sub-optimal', yet acceptable, to individual stakeholders). Thus, the scope is oriented to stakeholder groups, and is meant to help decision-makers, particularly in conflicting CE situations, where participants have clashing objectives; an aspect that has not been yet addressed previously in the CE literature.

The aim of this paper is to present a methodology whose underlying philosophy is to encourage cooperation between stakeholders within the decision-making process and, where cooperation of all is not possible, to demonstrate where decision-making is vulnerable to discordant views. Its starting point is the adoption of two underlying, well-evidenced principles: to be capable of realising the aspirational futures of a city (in this case, creating an effective CE), all urban stakeholders should ideally work jointly and collaborate effectively [24]; and to truly transform an urban area, a transdisciplinary approach must provide the foundation to solve city problems [25]. It also assumes *a priori* that stakeholders: are individually rational (are able to define objectives and appropriate actions that meet their own needs), have complete information, are willing to engage in a discourse with other actors, and are potentially willing to compromise on and accept compensation for their satisfaction levels (as long as their needs are sufficiently met); these assumptions being in accordance with GT principles. Finally, it assumes that it is possible to define a comprehensive stakeholder directory—those who have a vested interest and should be included as 'actors' in the GT process—that this set of actors remains comprehensive (i.e., no new actors will be introduced) and that all actors will continue to comply with GT principles (continue to engage in discourse and be willing to compromise, unless one or more wishes to withdraw from the process) during all later stages of the decision-making process.

This activity forms part of a larger decision-making process around a substantive change to (or an intervention in) the complex system-of-systems that make up cities and underpin civilised life. Citizens and those who govern them (city leaders) have aspirations for their place (visions, mission statements and suchlike), representing bottom-up and top-down perspectives, and these will almost certainly include many aspects of a CE, whether for economic, social, environmental or political reasons [24]. These need to be identified, articulated and disseminated to all stakeholders. The current operational paradigm (the systems that currently operate, e.g., often in accordance with a linear economy) needs to be understood and mapped, its current performance (in CE terms) established and a rigorous diagnosis of the problems of transformation to a CE carried out [25]. Only then can an engineered solution—a revised system operating in accordance with CE principles—be proposed [3]. This would inevitably attempt to take all relevant stakeholders' views into account while delivering a suite of benefits that meets the combined aspirations of the citizens and city leaders, while addressing national and global priorities [24]. It will equally explore how well the intervention is likely to function if the future context changes (to build in resilience), formulate alternative business models to secure the investment necessary to implement the intervention, and identify all of the forms of governance—formal (legislation, regulations, codes and standards) and informal (individual and societal attitudes and behaviours, social norms)—that would determine whether the intended benefits of the intervention would be likely to be delivered [3]. While this overarching set of methodologies is straightforward to define, one crucial question remains: will all of the actors involved—the individuals who will determine whether the intervention will work as intended—either positively enable it to work or allow it to work?

In addition to addressing this question, a crucial gap in knowledge addressed by this paper is that even though several researchers have studied CE implementation (e.g., [26–28]), and despite others recognising its relevance to the successful adoption of CE principles (e.g., [4–6]), cooperation between stakeholders (and its satisfactory achievement) has not yet been researched in terms of it being a key element for the CE transition. In dealing with these two primary goals, we illustrate and trial our thinking by considering adoption of CE principles in the waste management of Birmingham, UK's second city. The paper is organised as follows: Section 2 presents a six-step hybrid approach that integrates three existing methodologies, namely, SA, MCDA and GT; Section 3 describes the case study to which the proposed hybrid approach is applied (the first phase of decision-making, involving the five most influential actors); Section 4 discusses the results of its application; and Section 5 highlights the conclusions and potential future areas for research.

2. Methods and Materials

The hybrid approach consists of the six steps shown in Figure 1. Further details are provided in Sections 2.1–2.6.

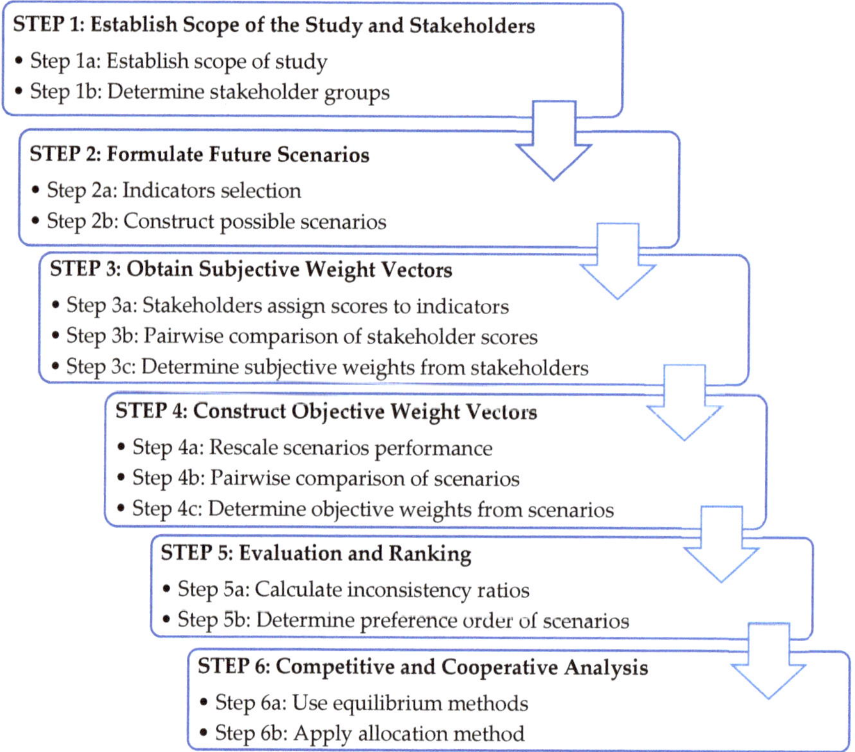

Figure 1. Flowchart summarising the methodology.

2.1. STEP 1: Establish Scope of the Study and Stakeholder Groups

To define and classify the stakeholder groups for consideration in the decision-making process of Municipal Solid Waste Management (MSWM), four steps are adapted from the 'Stakeholder Analysis Module' tool [29]:

1. Define the problem to study (Step 1a);
2. State the important elements that caused the problem (Step 1a);
3. List all stakeholders with an interest in the elements of the issue (Step 1b);
4. Remove any duplicated stakeholders (Step 1b).

2.1.1. Step 1a: Define Scope of Study

The scope of the study (i.e., problem to be solved) needs to be clearly stated and any influencing factors that caused the problem need to be considered. The objectives refer to specifying what is intended to be drawn out of the study, e.g., compare MSWM or EfW alternatives [9]. The case study application of this step is shown in Section 3.2.

2.1.2. Step 1b: Determine Stakeholder Groups

A stakeholder is defined as a group or an individual who is influencing or being influenced by (or both influencing and being influenced by) a set of decisions regarding a specific issue [16]. If the stakeholders are highly unlikely to cooperate, it might be thought that the possibilities of their being included in the decision-making process, or the effectiveness of the decision-making process itself, would be significantly reduced [30]. However, the proposed framework aims to encourage cooperation—it makes transparent to all actors the benefits of cooperation and the adverse consequences of failing to cooperate—and as such, advocates the inclusion of the opposing actors in order to shed light on their contradictory and/or contentious views. The methodology has been devised to increase the overall levels of satisfaction of all involved, and therefore has the potential to change the views of stakeholders (or actors in the decision-making process) who might initially adopt a contradictory stance in relation to the proposed (CE) intervention. This would, of course, be a rational response; irrational actors, who would disrupt any decision-making process, would be exposed as such as the methodology progresses and would find themselves isolated, and their views potentially excluded, as a result of the openness and transparency being brought to the decision-making. This might then result in them deciding to withdraw, or being asked by the collective to withdraw, and this would then become a matter of record when the final decision-making outcomes are disseminated.

2.2. *STEP 2: Formulate Future Scenarios*

It is now necessary to build future scenarios, of which their performance will be evaluated. Scenarios are usually constructed from the participant stakeholders, the available MSWM alternatives and the indicators to assess them [9]. Case study application of this step is shown in Section 3.3.

2.2.1. Step 2a: Indicators selection

Indicators must have characteristics considered to be appropriate to measuring the performance of CE scenarios. A good indicator should identify where you are and provide a pathway to where you want to be [31]; therefore, a unit of measure is required. The process of correctly identifying indicators relevant for CE assessment is based on:

- Valenzuela-Venegas et al. [32], which describes the process of selecting a range of sustainability indicators relevant to evaluating the sustainable performance of Eco-Industrial Parks (EIPs);
- Saidani et al. [33], which reviewed literature relevant to CE indicators; and
- Valenzuela-Venegas et al. [32] and Leach et al. [34] in which that the desirable properties for selecting indicators accurately are reported as:

    1. Understanding: be understood easily;
    2. Pragmatism: be easily measurable and data easily obtained;
    3. Relevance: be aligned with the goals and future of EIPs and businesses;
    4. Representative: enable the comparison of EIPs and allow for progress to be identified; and
    5. Multi-dimensional: evaluate one or more sustainability dimensions.

2.2.2. Step 2b: Construct Possible Scenarios

The objective of utilising MCDA is to evaluate—according to stakeholders' preferences—which is the best from a set of hypothetically built, by the researcher, future CE scenarios of MSWM in cities. While there are many ways to develop scenarios, the four Urban Futures scenario archetypes, which are themselves based on four of the six scenarios developed by the Global Scenario Group (their two scenarios involving societal breakdown were considered irrelevant to this exercise), are used because of their ability to provide diverse stakeholder engagement in futures; for more details refer to [35–37]. They have been re-interpreted to show the thinking for MSWM:

Scenario 1. Market Forces (MF). An extreme extension of the business-as-usual scenario, yet one in which social and environmental concerns are ignored completely. In terms of MSWM, this would likely mean that sustainability does not feature high up the agenda and waste is considered as a burden and typically something that costs rather than makes money and hence CE receives little investment or attention.

Scenario 2. Policy Reform (PR). This is based on strict enforcement of policy to achieve sustainability goals. In terms of MSWM, sustainability is likely to feature high on the agenda and strict policies for ever increasing charges for landfill and fines for waste production are likely to ensue.

Scenario 3. New Sustainability Paradigm (NSP). This scenario is shaped by widely accepted sustainable citizen values and behaviour. For MSWM, it is likely that citizens readily embrace CE principles and the governance systems support such implementation.

Scenario 4. Fortress World (FW). This scenario is characterised by highly polarised wealth distribution and wellbeing. In terms of MSWM, there are likely to be significant disparities in the way this issue is considered. The wealthy inside the fortress take care of their waste by pushing it out of the fortress causing negative consequences for those who lie outside—a "not in my backyard-ism" mentality ensues.

2.3. *STEP 3: Obtain Subjective Weight Vectors*

In this step, stakeholders are asked, through the use of a questionnaire, to rank the selected indicators (Step 3a). The researcher then pairwise compares their scores (Step 3b) in order that a subjective weighting is found (Step 3c). Case study application of this step is shown in Section 3.4.

2.3.1. Step 3a: Stakeholders Assign Scores to Indicators

The first part of the ranking process is to utilise a 'priority scale' [15] where stakeholders assign 'priority scores' to the indicators based on how important the indicators were based on a well-recognised 9-point 'Saaty scale'(9 being most relevant and 1 being least relevant). This technique was selected due to its ease of understanding to all stakeholders, but most importantly because it helps avoid any inconsistencies and facilitates pairwise comparisons.

2.3.2. Step 3b: Pairwise Comparison of Stakeholder Scores

Using the output of Step 3a, the pairwise comparison of stakeholder scores is performed. The matrix of pairwise comparisons consists of one plus the differences in the ranking values ($DV_{i,j}$) of each indicator assigned by each stakeholder, and they are calculated as follows. However, if the comparison results in a negative number, the value of $DV_{i,j}$ will be given by the reciprocal of one plus the absolute value of the differences:

$$DV_{i,j} = 1 + (R_i - C_j), \quad (1)$$

$$DV_{i,j} = \frac{1}{(1+|R_i - C_j|)}, \quad (2)$$

The matrix of pairwise comparisons should look like the following:

		Difference Values ($DV_{i,j}$)			
		C_j			
		1	**2**	...	**j**
	1	$DV_{1,1}$	$DV_{1,2}$	...	$DV_{1,j}$
$\mathbf{R_i}$	2	$DV_{2,1}$	$DV_{2,2}$	...	$DV_{2,j}$
	...	...	...	...	...
	i	$DV_{i,1}$	$DV_{i,2}$	...	$DV_{i,j}$
	Σ	$\Sigma\, C_1$	$\Sigma\, C_2$	...	$\Sigma\, C_j$

where, $DV_{i,j}$ is the difference between the values of the indicator in the row (R_i) minus the indicator in the column (C_j). The i and j subscripts are the row and column indicators, respectively. The sum of the columns in the bottom row (ΣC_j) is then used in the following Step 3c.

2.3.3. Step 3c: Determine Subjective Weights from Stakeholders

The calculations of the subjective indicator weights are to be performed using the well-known Analytical Hierarchy Process (AHP) technique. The method aims to produce weights for criteria, based on qualitative ranking data from decision-makers [38]. Using the output of Step 3b, to calculate the values of the normalised matrix, the following formula is used:

$$NV_{i,j} = \frac{DV_{i,j}}{\Sigma\, C_j}, \tag{3}$$

The indicator weight is calculated using the arithmetic mean of the normalised values for each row ($\overline{NV_i}$):

$$I_i = \overline{NV_i} = \frac{1}{n}\sum\nolimits_{i=1}^{n} NV_i \tag{4}$$

To obtain the exact weights, iterations must be performed until the new weights obtained do not change significantly from the value previously calculated. To do so, the set of weights must be multiplied by the original matrix of pairwise comparisons:

$$RV_i = \begin{bmatrix} \begin{bmatrix} I_1 \\ I_2 \\ \vdots \\ I_i \end{bmatrix} \begin{bmatrix} DV_{1,1} & DV_{1,2} & \cdots & DV_{1,j} \\ DV_{2,1} & DV_{2,2} & \cdots & DV_{2,j} \\ \vdots & \vdots & \cdots & \vdots \\ DV_{i,1} & DV_{i,2} & \cdots & DV_{i,j} \end{bmatrix} \end{bmatrix}, \tag{5}$$

$$\check{I}_i = \frac{RV_i}{\Sigma\, RV_i}, \tag{6}$$

The normalised matrix of pairwise comparisons should look like the following:

		Normalised Values ($NV_{i,j}$)				1st Weights	Iterations	Final Weights
		$\hat{C}_j$						
		1	**2**	**...**	**j**	$\bar{I}_i = NV_i$	RV_i	$\check{I}_i$
$\hat{R}_i$	1	$NV_{1,1}$	$NV_{1,2}$	...	$NV_{1,j}$	I_1	RV_1	$\check{I}_1$
	2	$NV_{2,1}$	$NV_{2,2}$	...	$NV_{2,j}$	I_2	RV_2	$\check{I}_2$
	...	...	...	...	...	...	...	...
	i	$NV_{i,1}$	$NV_{i,2}$	...	$NV_{i,j}$	I_i	RV_i	$\check{I}_i$
	Σ	$\Sigma \hat{C}_1$	$\Sigma \hat{C}_2$	...	$\Sigma \hat{C}_j$	$\Sigma I_i = 1$	ΣRV_i	$\Sigma \check{I}_i = 1$

where, $NV_{i,j}$, calculated using Equation (3), is the normalised value of the indicator in the row ($\hat{R}_i$) and the column ($\hat{C}_j$). The sum of the normalised columns ($\hat{C}_j$) must be equal to one. I_i is the (subjective) weight for indicator i, obtained through Equation (4). RV_i is the revised (weight) value for indicator i, calculated using Equation (5). $\check{I}_i$ is the final weight for indicator i, calculated using Equation (6).

2.4. STEP 4: Construct Objective Weight Vectors

As opposed to the subjective weight which refers to the stakeholders' ranking of indicators, the objective weight vectors refer to the CE scenarios weights, the rankings being built by the researcher. Case study application of this step is shown in Section 3.5.

2.4.1. Step 4a: Rescale Scenarios Performance

In order to prevent significant differences in the pairwise comparisons resulting in disproportional weights and the ranking of CE scenarios, the data needs to be rescaled. To do this, the 'priority scale' is considered again. However, the ranking of the CE scenarios is based on their expected performance for each CE indicator. The number of levels to use needs to be determined by setting a maximum allowed weight for a single scenario and by using Equation (7) ([15], p.2376).

$$1 = y + (y/c) \times (x-1),$$
$$c = (x-1)/((1/y)-1), \tag{7}$$

where y is the maximum allowed weight (i.e., the worst case where all scenarios are ranked in the lowest level except one, which is ranked as the topmost), x is the number of scenarios being compared, and c is the number of levels in the new scale. The data from the CE built scenarios must now be linearly rescaled using the following equations, depending whether the indicator's aim is to be maximised or minimised, respectively:

$$m' = \frac{m - a_{min}}{a_{max} - a_{min}} \times (d_{max} - d_{min}) + d_{min}, \tag{8}$$

$$m' = \frac{a_{min} - m}{a_{min} - a_{max}} \times (d_{min} - d_{max}) + d_{max}, \tag{9}$$

where a_{min} and a_{max} are the minimum and maximum range of the measurements, d_{min} and d_{max} are the minimum and maximum range of the intended target rescaling, $m \in [a_{min}, a_{max}]$ is the measurement to be rescaled, and m' is the rescaled measurement to the desired $[d_{min}, d_{max}]$ range values.

2.4.2. Step 4b: Pairwise Comparison of Scenarios

To continue using the AHP technique, it is necessary to now pairwise compare the CE scenarios in terms of their importance to each indicator [39]. This is done using Equations (1) and (2) according to the explanation provided in Section 2.3.2. However, the only difference is that a matrix of pairwise comparisons for all scenarios must now be elaborated for each indicator.

2.4.3. Step 4c: Determine objective weights for scenarios

Similarly to calculating the subjective weights of stakeholders, the objective weights for scenarios are calculated using the AHP technique (described in Section 2.3.3). Using again the outputs from the previous steps—rescaling and pairwise comparing scenarios based on their performance for each indicator—the utilisation of the 'priority scale' should facilitate its ranking and avoid inconsistencies. The main difference is that since multiple matrices of pairwise comparisons were elaborated (one for each indicator) in the previous step, the result will be a matrix of (objective) weights for scenarios, rather than a vector.

2.5. STEP 5: Evaluation and Ranking

After both the subjective and objective weights are calculated, it must be ensured that they are consistent. Thereafter, it is possible to determine the preferred order of the CE scenarios by the stakeholders. Case study application of this step is shown in Section 3.6.

2.5.1. Step 5a: Calculate Inconsistency Ratios

There must be consistency in the preference judgements of both indicators and CE scenarios. Having used the 'priority scale' [15], any inconsistency should have been avoided; however, this must be verified by calculating the Inconsistency Ratio (IR). Its maximum value must be below 10% for the judgements to be considered acceptable, otherwise they are considered to be purely random and unreliable. The IR is given by the following formulas [15,38,40]:

$$CI = \frac{\sum RV_i - k}{k - 1}, \tag{10}$$

$$IR = \frac{CI}{RI}, \tag{11}$$

where, CI is the consistency index; ΣRV_i is the sum of the revised (weight) values for indicator i (also known as the maximum Eigenvalue of the matrix) calculated using Equation (5); k is the matrix dimension; and RI is the random index based on a mean CI value for purely random matrices, given in Saaty ([41], p.966).

2.5.2. Step 5b: Determine Preference Order of Scenarios

To determine the preference order of CE scenarios, a preferability index is determined using the previously obtained weighting vectors for stakeholders and for CE scenarios. The preferability index is calculated by multiplying the (stakeholders) subjective vector weights (Step 3c) and the (scenarios) objective matrix weights (Step 4c), producing the preferability index vectors:

$$PI_x = \begin{bmatrix} I_{x,1} \\ I_{x,2} \\ \vdots \\ I_{x,i} \end{bmatrix} \begin{bmatrix} S_{1,1} & S_{1,2} & \cdots & S_{1,i} \\ S_{2,1} & S_{2,2} & \cdots & S_{2,i} \\ \vdots & \vdots & \cdots & \vdots \\ S_{m,1} & S_{m,2} & \cdots & S_{m,i} \end{bmatrix} = \begin{bmatrix} S_{1,1}I_{x,1} & + & S_{1,2}I_{x,2} & +\ldots+ & S_{1,n}I_{x,i} \\ S_{2,1}I_{x,1} & + & S_{2,2}I_{x,2} & +\ldots+ & S_{1,n}I_{x,i} \\ \vdots & + & \vdots & +\ldots+ & \vdots \\ S_{m,1}I_{x,1} & + & S_{m,2}I_{x,2} & +\ldots+ & S_{m,i}I_{x,i} \end{bmatrix} \tag{12}$$

where PI_x is the preferability indexes of stakeholder x, $I_{x,i} \in (0 \leq I_{x,i} \leq 1)$ is the stakeholder x subjective weight for indicator i, and $S_{m,i} \in (0 \leq S_{m,i} \leq 1)$ is the objective weight for CE scenario m and indicator i. These indexes show, on a scale from 0 to 1, how preferable the CE scenarios are to each stakeholder; their total when summed must be equal to 1.

2.6. STEP 6: Competitive and Cooperative Analysis

Once the preferred order of scenarios has been determined, a competitive and cooperative analysis is performed to enhance the possibilities of stakeholders cooperating (and continue cooperating) towards achieving their combined most preferred CE scenario. Case study application of this step is shown in Section 3.7.

2.6.1. Step 6a: Use Equilibrium Methods

Non-Cooperative Game Theory (NCGT) uses equilibrium methods to facilitate the most probable outcomes in interactive decision-making, in which the behaviour of stakeholders can be predicted since they prioritise their own objectives. The preferability indexes obtained from the last step are now utilised to construct the payoffs of each stakeholder; these represent the levels of satisfaction obtained from the combined preference selection of all stakeholders. The combined preferences are calculated as follows:

$$\varphi_s = \sum\nolimits_{x=1}^{s} (PI_{x,m}), \tag{13}$$

where φ_s is the combined preference of the CE scenarios (s) for stakeholders x, respectively, and $PI_{x,m}$ is the preferability index of stakeholder x for CE scenario m. In other words, φ_s is the sum of one preferability index on any CE scenario for each of the five stakeholders. The objective of calculating the preferences of stakeholders is to analyse them using GT techniques. In order to do so, the payoffs for the NCGT analysis need to be constructed:

$$\prod\nolimits_{s}^{x}, m = \varphi_s \times PI_{x,m}, \tag{14}$$

where $\prod_s^{x,m}$ is the payoff for stakeholder x if scenario m is their chosen alternative and given the combined preferences φ_s as calculated from Equation (13). Thus, m must coincide with the scenario in the combined preference for stakeholder x.

A Nash equilibrium [42] finds the combination of stakeholder preferences to scenarios which gives the highest possible level of satisfaction to each stakeholder. It also helps understand how stakeholders would not be motivated to change their scenario selection, i.e., such action would result in decreasing their (and other stakeholders) obtained satisfaction. If a state is stable under several equilibrium methods, it is highly likely to be the final resolution of a game [43]. Four different methods are used in the software to compute the Nash equilibrium. For the detailed methods, refer to the literature as follows:

1. Pure strategy equilibria [44,45];
2. Minimising Lyapunov function [46];
3. Global Newton tracing [47];
4. Solving systems of polynomial equations [48].

2.6.2. Step 6b: Apply Allocation Method

Once the equilibriums are obtained, applying allocation methods can help in preventing the potentially formed coalitions from being abandoned by the stakeholders in the future. To do so, the benefits of the participants must be assigned adequately. In this framework, the proportional gains of stakeholders from the payoffs of the selection of the combined preferences of CE scenarios above are called benefits. These are constructed as follows:

$$\beta_s^{x,m} = \overline{\prod\nolimits_{s}} \times \prod\nolimits_{s}^{x,m}, \tag{15}$$

where the arithmetic mean of the payoffs for all stakeholders in the selected coalition ($\overline{\prod_s}$) multiplied by the respective payoff of stakeholder *x* ($\prod_s^{x,m}$) is the benefit for stakeholder *x* ($\beta_s^{x,m}$). In other words, the benefit is the proportional gain of each stakeholder from the payoffs obtained for the combined preference of scenarios.

Cooperative Game Theory (CGT) is able to efficiently and equitably assign benefits and costs to a set of stakeholders (called coalitions) instead of optimising each of them separately [49]. The CGT allocation method used is the well-known Shapley value [50], which assumes participants agree to behave cooperatively and assigns each of them their marginal contribution to the coalition they join.

For the benefit allocation method, the data to use as input is the sum of benefits for each possible coalition ($\sum_{x=1}^{s} \beta_s^{x,m}$). The application of this method is performed using the R programming package 'GameTheory' to solve cooperative games [51]. The results for the Shapley value are obtained with R version 3.5.1 on Windows 10 version 1903. For a detailed description of the Shapley value method, refer to [50,51]. The Shapley value results are then compared with the original coalition total worth to reach the fairest distribution of the computed benefits. This final step yields a best allocation of "levels of satisfaction", and by tracing back the indicators which resulted as the highest weighted, focusing on increasing their performance would increase further the respondents' satisfaction. This provides the necessary evidence to increase the participants' satisfaction and encourage cooperation.

3. Case Study

3.1. General Overview

Tyseley Energy Park (TEP) in Birmingham, UK, is a renowned project aiming to adopt and develop sustainable energy generation technologies for the city at a time when its current EfW MSWM contract was due to expire (in 2019). Opportunities to embrace CE arise from such vast amounts of waste being at risk of remaining untreated. The land where the project is to be developed is privately owned and many business tenants are currently settled at the site. Other companies have private interests on settling businesses therein with IS potential and universities are interested in the research opportunities arising around renewable energies. The local government owns the incineration facilities and has a contract with a waste management operator to recover energy from a large portion of the MSW collected in the city, yet there exist general environmental and societal concerns by local inhabitants. Therefore, the local government and related companies have requested the help of consultancy services. This provides an opportunity for embedding CE principles in the development of TEP and there is no reason why this could not materialise if cooperation between stakeholders is achieved. Recent data on the city's MSWM [52] as depicted in Figure 2 shows that, even though Birmingham has one of the lowest landfill rates in the UK, two-thirds of the MSW is still treated via incineration with energy recovery. Despite annually producing 217 GWh of energy each year, this amount is equivalent to just over 1% of the city's total energy demand [52]. This casts doubt on the true circularity of the current treatment, and suggests that opportunities for improving the city's MSWM to be more aligned to CE principles could be generated and should be embraced.

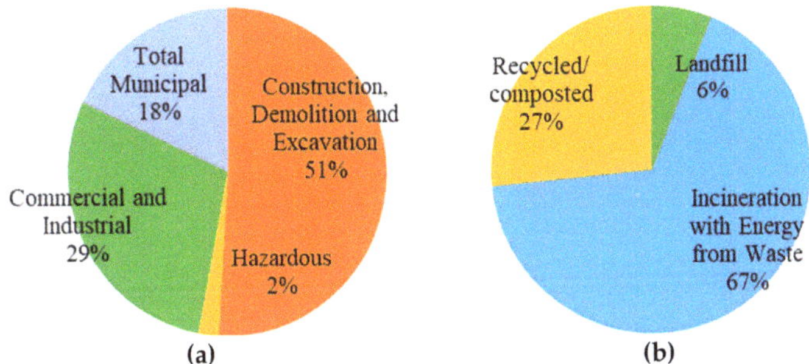

Figure 2. (**a**) Types of waste distribution for Birmingham; (**b**) Municipal Solid Waste (MSW) treatment (2013/14) for Birmingham (percentage amounts by weight).

3.2. STEP 1: Scope of study and stakeholders for TEP, Birmingham

3.2.1. Step 1a: Scope of the Study

Drawing on the description in the previous section, the scope of this study is to compare CE MSWM future scenarios for TEP, Birmingham. The primary influencing factor is that the current EfW plant in Tyseley is due to close, and CE opportunities arising from this have been widely discussed in meetings with the city stakeholders. The Birmingham case study is an academic exercise using a sub-set of real decision-makers from an ongoing process. The authors have performed solely as external observers, although the outcomes are being shared with the wider set of decision-makers.

3.2.2. Step 1b: Stakeholder Selection

The respondents in this study included a consultant in ISs, a university researcher on CE SMEs, a Tyseley ward local (community) representative, a steel sector company landowner in TEP, and a representative of Birmingham and Solihull Local Enterprise Partnership (representing local government). Drawing on the previous description of the problem and its elements, the respondents' identities and any information that could enable them to be traced back is not revealed and they were thus categorised in stakeholder groups as follows:

A. Companies—energy sector businesses in the TEP
B. Academic institutions
C. Local government
D. General public
E. Consultants—externals to the previous stakeholders who provide consultation services

3.3. STEP 2: Indicators selection and formulation of CE MSWM Scenarios

3.3.1. Step 2a: Indicators Selection

Applying Step 2 of the methodology (Section 2.2), the selected indicators for the case study are shown in Table 1. They have been selected through an extensive and thorough literature review. Their choice was based on being understandable, pragmatic, relevant, representative and able to assess a sustainability dimension (i.e., they all comply with the five recommended properties, see Section 2.2.1) and are deemed appropriate to the scale (i.e., city and eco-industrial park levels). In order to study the three acknowledged dimensions of CE and sustainability, three main indicator categories were selected (i.e., economic (indicators 1 to 3), environmental (indicators 4 to 7) and social (indicators 8 and 9). The appropriate units of measure and underpinning objective of the indicator (i.e., to minimise versus maximise the quantity) are shown. For example, indicator 1, Investment cost, is one of the most commonly used economic indicators [53]. It was set to be minimised, in line with Behera et al.'s [54] framework arguing that research performed in EIPs (as in TEP) is meant to be developed into a business (profit-led activities), thus aiming to reduce costs.

Table 1. Indicators adopted for the evaluation of Circular Economy (CE) waste management scenarios.

No.	Indicator	Unit	Description	Objective
1	Investment cost	£M/yr	It is the £million invested in a project.	Minimise
2	GVA [1] impact	£M/yr	It measures the total annual added production value at the end of the year.	Maximise
3	Payback	months	It indicates the time required for a project to recover the investment.	Minimise
4	Carbon emissions mitigation	CO_2 kt/yr	It reflects the amount of CO_2 emissions in kilotonnes that are reduced/saved yearly.	Maximise
5	MSW generation reduction	%	It measures the MSW generated in comparison to a previous year.	Maximise
6	Recycling rate of MSW	%	It measures the recycling rate of MSW, and the level of materials re-used and recycled in the city.	Maximise
7	Landfill rate of MSW	%	It measures the rate of MSW that is not diverted from disposal in the city.	Minimise
8	Jobs creation	#	It measures new jobs created per annum.	Maximise
9	Public awareness and satisfaction	%	It expresses the overview of opinions related to the MSWM system by the local population.	Maximise

[1] GVA—Gross Value Added.

3.3.2. Step 2b: Construct Possible Scenarios

The indicators are used to set performance levels within the future scenarios as shown in Table 2. (Note, the user could use more indicators than those shown here; an abridged set is used herein to aid understanding of the method). These values are not predictions but merely suggestions of likely future performance for the year 2030 drawn from data [55,56]. The supporting narrative for these values within each scenario has once again been removed for clarity.

Table 2. Future CE scenarios matrix for Municipal Solid Waste Management (MSWM) in Birmingham, UK.

Type	No.	Indicator	Scenarios				Unit
			MF	PR	NSP	FW	
Economic	1	Investment cost	46.4	53.8	60.6	43.1	£M/yr
	2	GVA impact	12.5	15.0	17.0	12.0	£M/yr
	3	Payback	180.0	300.0	360.0	240.0	months
Environmental	4	Carbon emissions mitigation	45.0	55.0	65.0	35.0	CO_2 kt/yr
	5	MSW generation reduction	7.0	8.5	10.0	3.0	%
	6	Recycling rate of MSW	31.5	40.0	50.0	30.0	%
	7	Landfill rate of MSW	6.0	2.5	1.0	5.0	%
Social	8	Jobs creation	1157.0	1469.0	1836.0	1101.0	#
	9	Public awareness and satisfaction	44.0	56.0	70.0	42.0	%

3.4. STEP 3: Subjective Weights for Indicators

3.4.1. Step 3a: Stakeholders Rank Indicators

In this step, the five critical stakeholder groups (shown as A to E in Table 3) were asked to rank the selected nine indicators (using a priority scale, see Section 2.2.1). Stakeholders were told not to use the same ranking number more than once; however, they were allowed to have more than one indicator with the same level of relevance. For example, stakeholder A and D gave indicators 1 and 3 the highest ranking of 9, likewise stakeholder C and E gave indicators 5, 6 and 7 an equal ranking of 3. None of the stakeholders deemed any indicator to be irrelevant.

Table 3. The indicators key on the left and stakeholders' responses on the right.

Type	No.	Indicator	Rank Value	Stakeholders' Responses				
				A	B	C	D	E
Economic	1	Investment cost	9	1, 3	5	2	1,3	2
	2	GVA impact	8	8	6, 7	8	4, 9	1
	3	Payback	7	9	3	1		3
Environmental	4	Carbon emissions mitigation	6	2	4	4		
	5	MSW generation reduction	5	4, 5, 6, 7	1	9		8, 9
	6	Recycling rate of MSW	4		8	3		4
	7	Landfill rate of MSW	3		9	5, 6, 7	2, 5, 6, 7	5, 6, 7
Social	8	Jobs creation	2		2		8	
	9	Public awareness and satisfaction	1					

3.4.2. Step 3b: Pairwise Comparison of Stakeholder Scores

The priority scale helped the stakeholders to simplify the process of pairwise comparing the selected indicators Table 4. A pairwise comparison matrix was produced from the filled-in priority scales Table 3 and by applying Equations (1) and (2). It is not possible to show the calculations for all pairwise comparisons of all 9 indicators and all 5 stakeholders, as there would need to be a total of $n \times (n-1)/2$ comparisons made. Hence, the set of pairwise comparisons for stakeholder B are shown in Table 4. For illustration, calculations associated with indicator 1 (i.e., Investment cost) are shown. For the first calculation, it can be seen from Table 3 (shaded in light grey) that stakeholder B gave Investment cost (R_1) a value of 5, and again Investment cost (C_1) a value of 5, hence the value of $DV_{1,1} = 1 + (R_1 - C_1) = 1 + (5 - 5) = 1$, likewise:

$$DV_{2,1} = \frac{1}{(1 + |R_2 - C_1|)} = \frac{1}{(1+|2-5|)} = 1/4$$

$$DV_{3,1} = 1 + (R_3 - C_1) = 1 + (7 - 5) = 3$$

$$DV_{4,1} = 1 + (R_4 - C_1) = 1 + (6 - 5) = 2$$

$$DV_{5,1} = 1 + (R_5 - C_1) = 1 + (9 - 5) = 5$$

$$DV_{6,1} = 1 + (R_6 - C_1) = 1 + (8 - 5) = 4$$

$$DV_{7,1} = 1 + (R_7 - C_1) = 1 + (8 - 5) = 4$$

$$DV_{8,1} = \frac{1}{(1 + |R_8 - C_1|)} = \frac{1}{(1 + |4 - 5|)} = 1/2$$

$$DV_{9,1} = \frac{1}{(1 + |R_9 - C_1|)} = \frac{1}{(1 + |3 - 5|)} = 1/3$$

Table 4. Example matrix of pairwise comparisons for stakeholder B.

			Column indicators								
			C_j								
			Investment cost	**GVA impact**	**Payback**	**Carbon emissions mitigation**	**MSW generation reduction**	**Recycling rate of MSW**	**Landfill rate of MSW**	**Jobs creation**	**Public awareness and satisfaction**
	Row Indicators		C_1	C_2	C_3	C_4	C_5	C_6	C_7	C_8	C_9
R_i	**Investment Cost**	R_1	1	4	1/3	1/2	1/5	1/4	1/4	2	3
	GVA Impact	R_2	1/4	1	1/6	1/5	1/8	1/7	1/7	1/3	1/2
	Payback	R_3	3	6	1	2	1/3	1/2	1/2	4	5
	Carbon Emissions Mitigation	R_4	2	5	1/2	1	1/4	1/3	1/3	3	4
	MSW Generation Reduction	R_5	5	8	3	4	1	2	2	6	7
	Recycling Rate of MSW	R_6	4	7	2	3	1/2	1	1	5	6
	Landfill Rate of MSW	R_7	4	7	2	3	1/2	1	1	5	6
	Jobs Creation	R_8	1/2	3	1/4	1/3	1/6	1/5	1/5	1	2
	Public Awareness and Satisfaction	R_9	1/3	2	1/5	1/4	1/7	1/6	1/6	1/2	1
	ΣC_j		20	43	9 4/9	14 2/7	3 2/9	5 3/5	5 3/5	26 5/6	34 1/2

It must be noted that the $DV_{i,j}$ values in the diagonal should be ones, because the comparisons between the indicator in the column minus the indicator in the row results in a subtraction of the exact same indicator plus one. Also, the values below the diagonal (shaded in dark grey) must mirror reciprocally those above it, because these comparisons are between the same indicators, but oppositely.

3.4.3. Step 3c: Determine Subjective Weights from Stakeholders

Using the outputs from Step 3b, the AHP technique is now applied. For illustration, the examples to be used are the shaded cells in Table 5. First, the normalised matrix of pairwise comparisons is calculated using Equation (3) and data from Table 4, the normalised value of the comparison between indicator 1 and indicator 1 is: $NV_{1,1,} = \frac{DV_{1,1}}{C_1} = \frac{1}{20} = 0.050$. The process is repeated to fill in the table. Second, the first weight for indicator 1 can be obtained using Equation (4), being the arithmetic mean of the row of indicator 1 ($\hat{R}_i$):

$$I_1 = \frac{1}{9}\sum\nolimits_{i=1}^{9} NV_i = \frac{1}{9}(0.500 + 0.93 + 0.035 + 0.035 + 0.062 + 0.045 + 0.075 + 0.087) = 0.0585$$

Table 5. Example normalised matrix of pairwise comparisons for stakeholder B.

Row Indicators		$\hat{C}_j$									1st Weights	Iteration 1	2nd Weights
		$\hat{C}_1$	$\hat{C}_2$	$\hat{C}_3$	$\hat{C}_4$	$\hat{C}_5$	$\hat{C}_6$	$\hat{C}_7$	$\hat{C}_8$	$\hat{C}_9$			
$\hat{R}_i$	$\hat{R}_1$	0.050	0.093	0.035	0.035	0.062	0.045	0.045	0.075	0.087	0.0585	0.5363	0.0568
	$\hat{R}_2$	0.012	0.023	0.018	0.014	0.039	0.026	0.026	0.012	0.014	0.0205	0.1873	0.0198
	$\hat{R}_3$	0.149	0.140	0.106	0.140	0.104	0.089	0.089	0.149	0.145	0.1235	1.1695	0.1240
	$\hat{R}_4$	0.100	0.116	0.053	0.070	0.078	0.060	0.060	0.112	0.116	0.0848	0.7912	0.0839
	$\hat{R}_5$	0.249	0.186	0.317	0.280	0.311	0.358	0.358	0.224	0.203	0.2761	2.6173	0.2774
	$\hat{R}_6$	0.199	0.163	0.212	0.210	0.155	0.179	0.179	0.186	0.174	0.1841	1.7553	0.1861
	$\hat{R}_7$	0.199	0.163	0.212	0.210	0.155	0.179	0.179	0.186	0.174	0.1841	1.7553	0.1861
	$\hat{R}_8$	0.025	0.070	0.026	0.023	0.052	0.036	0.036	0.037	0.058	0.0403	0.3661	0.0388
	$\hat{R}_9$	0.017	0.047	0.021	0.018	0.044	0.030	0.030	0.019	0.029	0.0282	0.2554	0.0271
$\Sigma\hat{C}_j$		1.000	1.000	1.000	1.000	1.000	1.000	1.000	1.000	1.000	1.0000	9.4337	1.0000

This 0.0585 value represents, on a scale from zero to one, how much stakeholder B considers indicator 1 to be worth in comparison with the rest of the indicators. To find the exact weights, several iterations must be performed. For illustration, only the first iteration is performed. Equation (5) is applied to multiply the 1st weight's vector by the original matrix of pairwise comparisons (Table 4). Thus, the revised (weight) value for indicator 1 (RV_1) is given by: $RV_1 = \sum_{i=1}^{9}(I_i \times DV_{1,j}) = I_1 \times DV_{1,1} + I_2 \times DV_{1,2} + I_3 \times DV_{1,3} + I_4 \times DV_{1,4} + I_5 \times DV_{1,5} + I_6 \times DV_{1,6} + I_7 \times DV_{1,7} + I_8 \times DV_{1,8} + I_9 \times DV_{1,9}$:

$$RV_1 = 0.0585 \times 1 + 0.0205 \times 4 + 0.1235 \times \tfrac{1}{3} + 0.0848 \times \tfrac{1}{2} + 0.2761 \times \tfrac{1}{5} + 0.1841 \times \tfrac{1}{4} + 0.1841 \times \tfrac{1}{4}$$
$$+0.0403 \times 2 + 0.0282 \times 3 = 0.5363$$

With the revised values, they must be normalised using Equation (6). Thus, this gives the second weight value for indicator 1: $\check{I}_1 = \frac{RV_1}{RV_i} = \frac{0.5363}{9.4337} = 0.0568$.

Note that the sum of the indicator weights must be equal to one. The process iterates until the newly calculated weights are not significantly different to those previously computed. The final subjective weight vectors for all five stakeholders are shown in Table 6. The bottom row is the IR calculated for the final weights; this will be explained in Section 3.6.1. The calculations were performed with the aid of an adapted Excel template [57].

Table 6. Subjective weight vectors for stakeholders.

Type	No.	Indicator	Stakeholders				
			A	B	C	D	E
Economic	1	Investment cost	0.2465	0.0564	0.1540	0.2518	0.2282
	2	GVA impact	0.0694	0.0200	0.3132	0.0335	0.3194
	3	Payback	0.2465	0.1232	0.0465	0.2518	0.1614
Environmental	4	Carbon emissions mitigation	0.0412	0.0832	0.1052	0.1701	0.0483
	5	MSW generation reduction	0.0412	0.2799	0.0296	0.0335	0.0304
	6	Recycling rate of MSW	0.0412	0.1858	0.0296	0.0335	0.0304
	7	Landfill rate of MSW	0.0412	0.1858	0.0296	0.0335	0.0304
Social	8	Jobs creation	0.1642	0.0386	0.2217	0.0220	0.0756
	9	Public awareness and satisfaction	0.1087	0.0271	0.0707	0.1701	0.0756
		IR	1.1717%	2.4673%	2.6644%	1.3273%	2.5760%

3.5. STEP 4: Objective Weights for CE MSWM Scenarios

3.5.1. Step 4a: Rescale Scenarios Performance

Before being able to pairwise compare the scenarios, their suggested performances need to be rescaled into the 'priority scale' range. However, first, the maximum number of rankings needs to be determined. By substituting in Equation (7): four being the number of scenarios ("x"), and 0.5 the maximum allowed weight for a single scenario ("y"). Thus, three is the maximum times that a scenario is allowed to be more important than another. The number of levels to use ("c") is three (the top 9 to 7 from the 'priority scale'):

$$c = (4-1)/((1/0.5)-1) = 3$$

Using Equations (8) and (9), the data for CE scenarios in Table 2 are rescaled and shown in Table 7. For example, for indicator 1, Investment cost, to rescale the scenario values in Table 2 the formula to use is Equation (9) because the objective of the indicator is to be minimised as follows:

$$\mathrm{MF'}_1 = \frac{43.1 - 46.4}{43.1 - 60.6} \times (7-9) + 9 = 8.6$$

$$\mathrm{PR'}_1 = \frac{43.1 - 53.8}{43.1 - 60.6} \times (7-9) + 9 = 7.8$$

$$\mathrm{NSP'}_1 = \frac{43.1 - 60.6}{43.1 - 60.6} \times (7-9) + 9 = 7.0$$

$$\mathrm{FW'}_1 = \frac{43.1 - 43.1}{43.1 - 60.6} \times (7-9) + 9 = 7.0$$

Table 7. Rescaled suggested performance of CE scenarios.

	Indicators								
	Investment Cost	GVA impact	Payback	Carbon Emissions Mitigation	MSW Generation Reduction	Recycling Rate of MSW	Landfill Rate of MSW	Jobs Creation	Public Awareness and Satisfaction
Scenarios	C_1	C_2	C_3	C_4	C_5	C_6	C_7	C_8	C_9
MF	8.6	7.2	9.0	7.7	8.1	7.2	7.0	7.2	7.1
PR	7.8	8.2	7.7	8.3	8.6	8.0	8.4	8.0	8.0
NSP	7.0	9.0	7.0	9.0	9.0	9.0	9.0	9.0	9.0
FW	9.0	7.0	8.3	7.0	7.0	7.0	7.4	7.0	7.0

Values in between the levels (top 9 to 7 from the 'priority scale') are used, thus the first decimal rounded figure is used for further AHP calculations.

3.5.2. Step 4b: Pairwise Comparison of Scenarios

Similar to pairwise comparison of the stakeholders rankings, the scenarios rescaled performances are pairwise compared using the 'priority scale' according to each indicator individually. This process used the output from the previous step Table 8 to create the matrices for each indicator. For illustration purposes, only the matrix for indicator 1, Investment cost, is presented in Table 8. The rest of the indicators have their own matrix of pairwise comparisons; while they are not presented in this paper, they follow the same calculating procedure. For the first calculation in Table 8, it can be seen from Table 7 (shaded in light grey) that for indicator 1, Investment cost (C_1) and scenario MF, the rescaled value of the MF scenario is 8.6, and again the rescaled value of the MF scenario is 8.6, hence the difference value is:

$$DV_{MF,MF} = 1 + (MF_1 - MF_1) = 1 + (8.6 - 8.6) = 1.000, \text{ likewise :}$$

$$DV_{PR,MF} = \frac{1}{(1 + |PR_1 - MF_1|)} = \frac{1}{(1 + |7.8 - 8.6|)} = 0.556$$

$$DV_{NSP,MF} = \frac{1}{(1 + |NSP_1 - MF_1|)} = \frac{1}{(1 + |7.0 - 8.6|)} = 0.385$$

$$DV_{FW,MF} = 1 + (FW_1 - MF_1) = 1 + (9.0 - 8.6) = 1.400$$

Table 8. Example matrix of pairwise comparisons of scenarios for indicator 1, Investment cost.

Indicator 1: Investment Cost		**Column Scenarios** C_j			
Row scenarios		**MF**	**PR**	**NSP**	**FW**
R_i	**MF**	1.000	1.800	2.600	0.714
	PR	0.556	1.000	1.800	0.455
	NSP	0.385	0.556	1.000	0.333
	FW	1.400	2.200	3.000	1.000
	ΣC_j	3.340	5.556	8.400	2.502

3.5.3. Step 4c: Determine Objective Weights for Scenarios

The AHP technique is similarly applied to determine the objective weights of scenarios. For illustration, the examples to be used are the shaded cells in Table 9. First, the normalised matrix of pairwise comparisons is calculated using Equation (3) and data from Table 8. The normalised value of the comparison for indicator 1 between scenario MF and scenario MF is: $NV_{MF,MF} = \frac{DV_{MF,MF}}{\sum C_{MF}} = \frac{1}{3.340} = 0.299$. The process is repeated to fill in the table. Second, the first weight for indicator 1 and scenario MF can be obtained using Equation (4), being the arithmetic mean of the row of scenario MF:

$$S_{MF,1} = \frac{1}{4}\sum_{i=1}^{4} NV_{MF,i} = \frac{1}{4}(0.299 + 0.324 + 0.310 + 0.285) = 0.3046$$

Table 9. Example normalised matrix of pairwise comparisons of CE scenarios for indicator 1.

Row Scenarios		$\hat{C}_j$				**1st Weights**	**Iteration 1**	**2nd Weights**
		$\hat{C}_{MF}$	$\hat{C}_{PR}$	$\hat{C}_{NSP}$	$\hat{C}_{FW}$			
$\hat{R}_i$	$\hat{R}_{MF}$	0.299	0.324	0.310	0.285	0.3046	1.2231	0.3048
	$\hat{R}_{PR}$	0.166	0.180	0.214	0.182	0.1856	0.7438	0.1853
	$\hat{R}_{NSP}$	0.115	0.100	0.119	0.133	0.1169	0.4681	0.1166
	$\hat{R}_{FW}$	0.419	0.396	0.357	0.400	0.3930	1.5782	0.3933
$\Sigma\hat{C}_j$		1.000	1.000	1.000	1.000	1.0000	4.0132	1.0000

To find the exact weights, several iterations must be performed. For illustration, only the first iteration is presented. Equation (5) is applied to multiply the 1st weight's vector by the original matrix of pairwise comparisons (Table 8). Thus, the revised weight value for scenario MF and indicator 1 is:

$$RV_{MF,1} = \sum_{i=4}^{4} (S_{i,1} \times DV_{MF,i}) = S_{MF,1} \times DV_{MF,MF} + S_{PR,1} \times DV_{MF,PR} + S_{NSP,1} \times DV_{MF,NSP} + S_{FW,1} \times DV_{MF,FW}$$

$$RV_{MF,1} = 0.3046 \times 1.000 + 0.1856 \times 1.800 + 0.1169 \times 2.600 + 0.3930 \times 0.714 = 1.2231$$

The revised values must be normalised using Equation (6). Thus, this gives the second weight value for scenario MF and indicator 1: $\check{S}_{MF,1} = \frac{RV_{MF,1}}{\sum RV_i} = \frac{1.2231}{4.0132} = 0.3048$.

Table 10 presents the final objective weight matrix for all four CE scenarios and all nine indicators after the necessary iterations. Each weight represents the performance of the scenario in comparison to the rest of the scenarios for each indicator individually. The rightmost column is the IR calculated for the final weights (this is addressed in Section 3.6.1).

Table 10. Objective weight matrix for the CE scenarios.

		Scenarios				
	Indicators [1]	**MF**	**PR**	**NSP**	**FW**	**IR**
1	**Investment Cost**	0.3047	0.1853	0.1167	0.3933	0.3859%
2	**GVA Impact**	0.1509	0.2781	0.4387	0.1323	0.4302%
3	**Payback**	0.4182	0.1860	0.1221	0.2736	0.4564%
4	**Carbon Emissions Mitigation**	0.1860	0.2736	0.4182	0.1221	0.4564%
5	**MSW Generation Reduction**	0.2108	0.2934	0.3825	0.1132	0.3526%
6	**Recycling Rate of MSW**	0.1552	0.2580	0.4511	0.1357	0.4420%
7	**Landfill Rate of MSW**	0.1253	0.2939	0.4200	0.1607	0.4018%
8	**Jobs Creation**	0.1552	0.2580	0.4511	0.1357	0.4420%
9	**Public Awareness and Satisfaction**	0.1479	0.2606	0.4534	0.1381	0.4466%
	Total	1.8542	2.2871	3.2539	1.6049	

[1] Note that the columns and rows of the matrix have been transposed for presentation purposes.

3.6. STEP 5: Ranking Order of the CE Scenarios

3.6.1. Step 5a: Calculate Inconsistency Ratios

The use of the 'priority scale' has ensured that there are no inconsistencies in the rankings. However, this must be verified using Equations (10) and (11). For illustration, the IR of the scenarios' comparison for indicator 1 (IR_1) will be explained. The final iteration resulted in the maximum Eigenvector or sum of revised values being $\sum RV_1 = 4.0104$ (very similar to that in the first iteration in Table 9; the matrix dimension is 4; and the RI for a matrix of such dimension is 0.89 (obtained from [41]). Thus, the IR_1 is as follows:

$$CI_1 = \frac{\sum RV_1 - k}{k - 1} = \frac{4.0104 - 4}{4 - 1}$$

$$IR_1 = \frac{CI_1}{RI_4} = \frac{0.0035}{0.89} = 0.3859\%$$

All the IR values for both the final subjective and objective weights were presented above in the lowest row and the rightmost column in Tables 6 and 10, respectively. The IR values are well below the maximum acceptable value (IR < 10%).

3.6.2. Step 5b: Determine preference order of scenarios

To reveal the preference order of the CE scenarios for each stakeholder, the preferability indexes are calculated. Using Equation (12), the indexes for each stakeholder are shown in Table 11, and below them, their ranking order. The sum of the preferability indexes must be 1 for each stakeholder. For example, the preferability index for stakeholder B for the MF scenario ($PI_{B,MF}$) (shaded in light grey in Table 11) is the sum of all the products of each subjective weight ($I_{x,n}$) (shaded in Table 6 multiplied by the objective weight $S_{MF,n}$ (shaded in Table 10) for the MF scenario and indicator n:

$$PI_{B,MF} = \sum_{n=1}^{9} (I_{B,n} \times S_{MF,n}),$$

$$PI_{B,MF} = I_{B,1} \times S_{MF,1} + I_{B,2} \times S_{MF,2} + I_{B,3} \times S_{MF,3} + I_{B,4} \times S_{MF,4} + I_{B,5} \times S_{MF,5} + I_{B,6} \times S_{MF,6} + I_{B,7} \times S_{MF,7} + I_{B,8} \times S_{MF,8} + I_{B,9} \times S_{MF,9}$$

$$PI_{B,MF} = 0.0564 \times 0.3047 + 0.0200 \times 0.1509 + 0.1232 \times 0.4182 + 0.0832 \times 0.1860 + 0.2799 \times 0.2108 + 0.1858 \times 0.1552 + 0.1858 \times 0.1253 + 0.0386 \times 0.1552 + 0.0271 \times 0.1479 = 0.208$$

Table 11. Preferability indexes and ranking for the CE scenarios for each stakeholder.

Scenarios	Stakeholders				
	A	B	C	D	E
MF	0.258 (2nd)	0.208 (3rd)	0.193 (3rd)	0.264 (2nd)	0.232 (3rd)
PR	0.228 (4th)	0.263 (2nd)	0.254 (2nd)	0.228 (4th)	0.239 (2nd)
NSP	0.281 (1st)	0.364 (1st)	0.374 (1st)	0.275 (1st)	0.313 (1st)
FW	0.233 (3rd)	0.164 (4th)	0.180 (4th)	0.233 (3rd)	0.215 (4th)

3.7. STEP 6: GT Analysis

3.7.1. Step 6a: Use equilibrium Methods

Using the preferences of scenarios above, all their possible combinations for the four CE scenarios and the five stakeholders are calculated (4^5 = 1024 combinations). For example, using Equation (13) and data from Table 11, the combined preference of stakeholder A to MF, B to FW, C to NSP, D to FW and E to PR is given by:

$$\varphi_{MF,FW,NSP,FW,PR} = \sum PI_{A,MF} + PI_{B,FW} + PI_{C,NSP} + PI_{D,FW} + PI_{E,PR}$$

$$\varphi_{MF,FW,NSP,FW,PR} = \sum 0.258 + 0.164 + 0.374 + 0.233 + 0.239 = 1.270$$

The total number of payoffs to be calculated are five (one for each stakeholder) per combined preference (i.e., 1,024 × 5 = 5120 payoffs). For example, from Equation (14), the payoffs vector for the combined preference in the example above is:

$$\prod\nolimits_{MF,FW,NSP,FW,PR}^{A,MF} = \varphi_{MF,FW,NSP,FW,PR} \times PI_{A,MF} = 1.270 \times 0.258 = 0.328$$

$$\prod\nolimits_{MF,FW,NSP,FW,PR}^{B,FW} = \varphi_{MF,FW,NSP,FW,PR} \times PI_{B,FW} = 1.270 \times 0.164 = 0.209$$

$$\prod\nolimits_{MF,FW,NSP,FW,PR}^{C,NSP} = \varphi_{MF,FW,NSP,FW,PR} \times PI_{C,NSP} = 1.270 \times 0.374 = 0.475$$

$$\prod\nolimits_{MF,FW,NSP,FW,PR}^{D,FW} = \varphi_{MF,FW,NSP,FW,PR} \times PI_{D,FW} = 1.270 \times 0.233 = 0.296$$

$$\prod\nolimits_{MF,FW,NSP,FW,PR}^{E,PR} = \varphi_{MF,FW,NSP,FW,PR} \times PI_{E,PR} = 1.270 \times 0.239 = 0.304$$

$$\prod_{MF,FW,NSP,FW,PR} = [0.328, 0.209, 0.475, 0.296, 0.304],$$

The NCGT equilibrium analysis of payoffs for the stakeholders is performed using the open-access software Gambit (Version 15.1.1, The Gambit Project, Norwich, UK). These payoffs are a representation of the level of satisfaction obtained by each stakeholder for that specific combination of alternative scenarios. Thus, it is of great relevance to uncover the set or the single combination of scenarios which brings an equilibrium to the interactive decision-making. This means finding the highest possible satisfaction to each stakeholder without decreasing that obtained by others. The results in Table 12 show that, as initially expected, the calculated Nash equilibrium for the majority of the methods is when all stakeholders select the same NSP scenario ($\prod_{NSP,NSP.NSP,NSP,NSP}$). The payoffs are shown in the row below the scenarios.

Table 12. Nash Equilibriums for the combined preferences of stakeholders.

Equilibrium Methods	Nash Equilibriums Found	Stakeholders				
		A	B	C	D	E
Pure strategy Equilibria	1	NSP 0.453	NSP 0.585	NSP 0.602	NSP 0.442	NSP 0.504
Minimising Lyapunov Function	0	- -	- -	- -	- -	- -
Global Newton Tracing	1	NSP 0.453	NSP 0.585	NSP 0.602	NSP 0.442	NSP 0.504
Solving Systems of Polynomial Equations	1	NSP 0.453	NSP 0.585	NSP 0.602	NSP 0.442	NSP 0.504

3.7.2. Step 6b: Apply Allocation Method

Once the Nash equilibrium is identified, the next step is to analyse how stakeholders should arrange their satisfaction levels to prevent them from abandoning their (presumably) cooperative behaviour. This is done by applying CGT allocation methods, which aim to distribute benefits fairly to stakeholders in pre-emptive coalitions to protect cooperation.

For example, using Equation (15), the benefits of coalition ABCDE (where all stakeholders cooperate and are allocated benefits) for the equilibrium result where all stakeholders select the NSP scenario ($\prod_{NSP,NSP.NSP,NSP,NSP}$) is as follows:

$$\prod_{NSP,NSP,NSP,NSP,NSP} = [0.453, 0.585, 0.602, 0.442, 0.504]$$

$$\overline{\prod}_{NSP,NSP,NSP,NSP,NSP} = 0.517$$

$$\beta^{A,NSP}_{NSP,NSP,NSP,NSP,NSP} = \sum \overline{\prod}_{NSP,NSP,NSP,NSP,NSP} \times \prod{}^{A,NSP}_{NSP,NSP,NSP,NSP,NSP,NSP} = 0.517 \times 0.453 = 0.234$$

$$\beta^{B,NSP}_{NSP,NSP,NSP,NSP,NSP} = \sum \overline{\prod}_{NSP,NSP,NSP,NSP,NSP} \times \prod{}^{B,NSP}_{NSP,NSP,NSP,NSP,NSP,NSP} = 0.517 \times 0.585 = 0.302$$

$$\beta^{C,NSP}_{NSP,NSP,NSP,NSP,NSP} = \sum \overline{\prod}_{NSP,NSP,NSP,NSP,NSP} \times \prod{}^{C,NSP}_{NSP,NSP,NSP,NSP,NSP,NSP} = 0.517 \times 0.602 = 0.311$$

$$\beta^{D,NSP}_{NSP,NSP,NSP,NSP,NSP} = \sum \overline{\prod}_{NSP,NSP,NSP,NSP,NSP} \times \prod{}^{D,NSP}_{NSP,NSP,NSP,NSP,NSP,NSP} = 0.517 \times 0.442 = 0.229$$

$$\beta^{E,NSP}_{NSP,NSP,NSP,NSP,NSP} = \sum \overline{\prod}_{NSP,NSP,NSP,NSP,NSP} \times \prod{}^{E,NSP}_{NSP,NSP,NSP,NSP,NSP,NSP} = 0.517 \times 0.504 = 0.260$$

$$\beta_{NSP,NSP,NSP,NSP,NSP} = [0.234, 0.302, 0.311, 0.229, 0.260]$$

The rightmost column in Table 13 shows the sum of the benefits for each coalition, or the total worth of the coalition. It is essential to analyse every possible coalition that can be formed, i.e., from stakeholders working individually to the case where all of them behave cooperatively. Table 13 presents all the possible coalitions to be formed and their $\beta_{NSP,NSP,NSP,NSP,NSP}$ benefits.

Table 13. Coalitions of stakeholders for obtained Nash equilibrium.

Coalition Name	Payoffs $\Pi_{NSP,NSP,NSP,NSP,NSP}$						Benefits $\beta_{NSP,NSP,NSP,NSP,NSP}$					$\Sigma\beta$
	A	B	C	D	E	-	A	B	C	D	E	
	NSP	NSP	NSP	NSP	NSP	Π	NSP	NSP	NSP	NSP	NSP	
A	0.453	0	0	0	0	0.453	0.205	0	0	0	0	0.205
B	0	0.585	0	0	0	0.585	0	0.342	0	0	0	0.342
C	0	0	0.602	0	0	0.602	0	0	0.362	0	0	0.362
D	0	0	0	0.442	0	0.442	0	0	0	0.196	0	0.196
E	0	0	0	0	0.504	0.504	0	0	0	0	0.254	0.254
AB	0.453	0.585	0	0	0	0.519	0.235	0.303	0	0	0	0.538
AC	0.453	0	0.602	0	0	0.527	0.239	0	0.317	0	0	0.556
AD	0.453	0	0	0.442	0	0.447	0.202	0	0	0.198	0	0.400
AE	0.453	0	0	0	0.504	0.478	0.216	0	0	0	0.241	0.457
BC	0	0.585	0.602	0	0	0.593	0	0.347	0.357	0	0	0.704
BD	0	0.585	0	0.442	0	0.514	0	0.300	0	0.227	0	0.527
BE	0	0.585	0	0	0.504	0.544	0	0.318	0	0	0.274	0.592
CD	0	0	0.602	0.442	0	0.522	0	0	0.314	0.231	0	0.545
CE	0	0	0.602	0	0.504	0.553	0	0	0.332	0	0.278	0.610
DE	0	0	0	0.442	0.504	0.473	0	0	0	0.209	0.238	0.447
ABC	0.453	0.585	0.602	0	0	0.546	0.247	0.320	0.329	0	0	0.896
ABD	0.453	0.585	0	0.442	0	0.493	0.223	0.289	0	0.218	0	0.730
ABE	0.453	0.585	0	0	0.504	0.514	0.232	0.300	0	0	0.259	0.791
ACD	0.453	0	0.602	0.442	0	0.499	0.226	0	0.300	0.221	0	0.747
ACE	0.453	0	0.602	0	0.504	0.519	0.235	0	0.312	0	0.262	0.809
ADE	0.453	0	0	0.442	0.504	0.466	0.211	0	0	0.206	0.235	0.652
BCD	0	0.585	0.602	0.442	0	0.543	0	0.318	0.327	0.240	0	0.885
BCE	0	0.585	0.602	0	0.504	0.563	0	0.330	0.339	0	0.284	0.953
BDE	0	0.585	0	0.442	0.504	0.510	0	0.298	0	0.226	0.257	0.781
CDE	0	0	0.602	0.442	0.504	0.516	0	0	0.310	0.228	0.260	0.798
ABCD	0.453	0.585	0.602	0.442	0	0.520	0.235	0.304	0.313	0.230	0	1.082
ABCE	0.453	0.585	0.602	0	0.504	0.536	0.242	0.313	0.322	0	0.270	1.147
ABDE	0.453	0.585	0	0.442	0.504	0.496	0.224	0.290	0	0.219	0.250	0.983
ACDE	0.453	0	0.602	0.442	0.504	0.500	0.226	0	0.301	0.221	0.252	1.000
BCDE	0	0.585	0.602	0.442	0.504	0.533	0	0.312	0.321	0.236	0.269	1.138
ABCDE	0.453	0.585	0.602	0.442	0.504	0.517	0.234	0.302	0.311	0.229	0.260	1.336

The Shapley value results are shown in Table 14. The fairest allocation of benefits corresponds to the five-stakeholder coalition, in other words, when they decide to cooperate and remain in the previously agreed alliance. The letter below each allocated benefit in Table 14 indicates whether it should stay the same (S), increase (I) or decrease (D) compared to the originally claimed benefits in the coalition in Table 13. The sum of the newly proposed distribution must be equal to the total worth of the original coalition. The implications of these results are discussed in Section 4.

Table 14. Shapley value results for the benefits of stakeholders.

Allocation Method	Benefits $\beta_{NSP,NSP,NSP,NSP,NSP}$					
	Stakeholders					$\Sigma\beta$
	A	B	C	D	E	
Shapley value	0.200 (D)	0.337 (I)	0.355 (I)	0.191 (D)	0.253 (D)	1.336

4. Discussion

The total subjective weights for the indicators is depicted in Figure 3, in which it can be seen that the environmental indicators have resulted as of the least concern. Previous findings [9] indicate that industry stakeholders prefer economic indicators, whilst municipalities consider environmental indicators as being more important. For this case study, recycling and landfill rates of MSW have yielded the lowest weighted values, whereas the economic indicators resulted in the highest weight values. Academic institutions are the most concerned with environmental indicators. The slightly higher value for the reduction of carbon emissions might be related to the fact that Birmingham is committed to reduce its carbon footprint by 60% by 2027 [55,58,59].

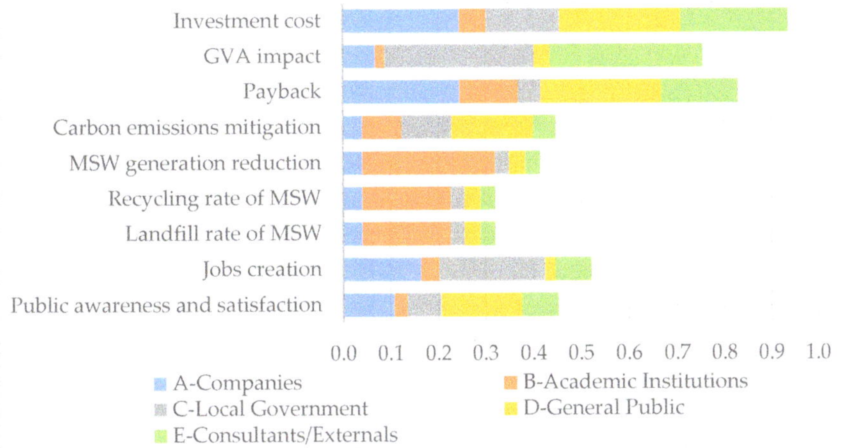

Figure 3. Subjective weight values for stakeholders.

An unexpected finding is the low subjective weight for the Jobs creation indicator for the General Public stakeholder (D). As mentioned by them during the interview: '(...) it's not just about jobs creation, we need skilled jobs in the area, not simple jobs (...)'. Conversely, the most important indicators for the rest of the stakeholders were Investment cost and Payback. This reinforces the initial expectations that the stakeholders' conflicting viewpoints might be a barrier to cooperate and thus reach the optimal scenario. The scenario that scored highest was NSP, followed by PR; MF was ranked third and the FW scenario resulted as the lowest ranked of all (Figure 4).

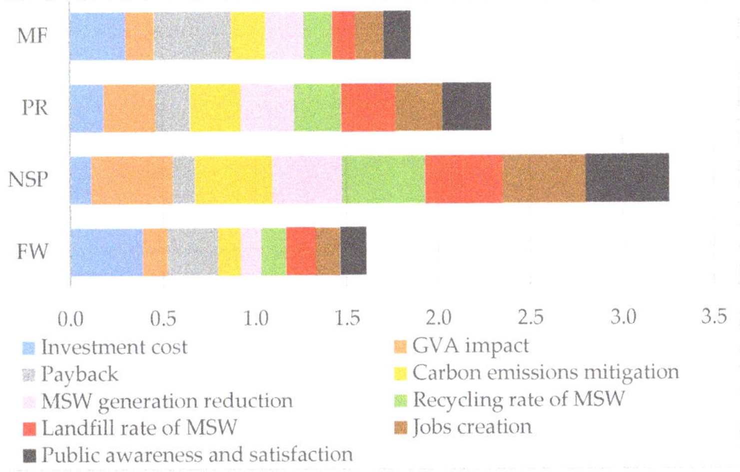

Figure 4. Objective weight values for scenarios.

NSP also resulted as the most preferred scenario for all stakeholders (Figure 5). This is in line with previous observations where 70% or their interviewed stakeholders ranked highest the most sustainable performing composting plant site alternative [15]. However, the second most preferred scenario varied between stakeholders. For example: stakeholders A (Companies) and D (General Public) ranked MF, FW and PR in second, third and fourth places, respectively. This means that they prefer a business-as-usual and a breakdown scenario over a strong policy implementation. In contrast, stakeholders B, C and E ranked PR, MF and FW in decreasing order. This suggested, before the GT analysis, that stakeholders having NSP as their most preferred scenario would be willing to work jointly towards it. However, it does not necessarily mean that their priorities are aligned, and that cooperation would occur naturally.

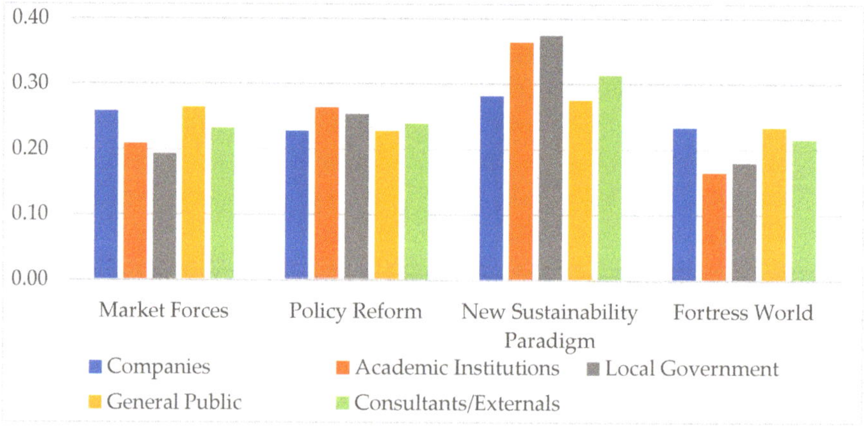

Figure 5. Preferability indexes for the CE scenarios for each stakeholder.

After the preferences of the stakeholders to the CE scenarios were revealed, the NCGT analysis reported that, as expected, stakeholders achieve their maximum levels of satisfaction (payoff) when all four of them select the NSP scenario, meaning this is a Nash equilibrium. If any of the participants were to deviate from this selection unilaterally, not only would that result in a decrease for them, but it would also result in a decrease to the rest of the stakeholders. This combined set of preferences ($\prod_{NSP,NSP,NSP,NSP,NSP}$) was then used to calculate the benefits system for the stakeholders ($\beta_{NSP,NSP,NSP,NSP,NSP}$) to enable the CGT analysis to be carried out.

The first row in Table 15 indicates the benefits each stakeholder would obtain separately. This implies there is no cooperation, and thus why there is no addition in the rightmost column. The second row shows the benefits obtained by each stakeholder if they all join a coalition and cooperate, with the letters below each entry showing how the benefits obtained compared to the previous benefits—some stakeholders (A, D and E) can increase their benefits whilst the rest (B and C) exhibit a decrease. In the bottom row, the Shapley value assigns benefits differently, with the letters below indicating how this new allocation compares with the previous case in which all stakeholders cooperate (ABCDE).

Table 15. Coalition distributions of benefits for stakeholders.

Coalition	Benefits $\beta_{NSP,NSP,NSP,NSP,NSP}$					$\Sigma\beta$
	Stakeholders					
	A	B	C	D	E	
Independently	0.205	0.342	0.362	0.196	0.254	
ABCDE	0.234 (I)	0.302 (D)	0.311 (D)	0.229 (I)	0.260 (I)	1.336
Shapley value	0.200 (D)	0.337 (I)	0.355 (I)	0.191 (D)	0.253 (D)	1.336

The Shapley value results in lower assignations to all stakeholders than if they work on their own (i.e., when a single stakeholder is considered in five different coalitions, shown in the first row in Table 15). Compared to the values for the ABCDE coalition, the benefit allocation for stakeholders B and C is suggested to increase, because according to the Shapley value definition, their allocation is influenced by their contribution to the coalition. In other words, it is a representation of their bargaining power and, as shown in their independent and ABCDE values, their contributions are the highest. Likewise, Figure 6 helps to visualise these comparisons and shows how the Shapley value is assigning the minimum satisfaction to prevent them from abandoning the coalition.

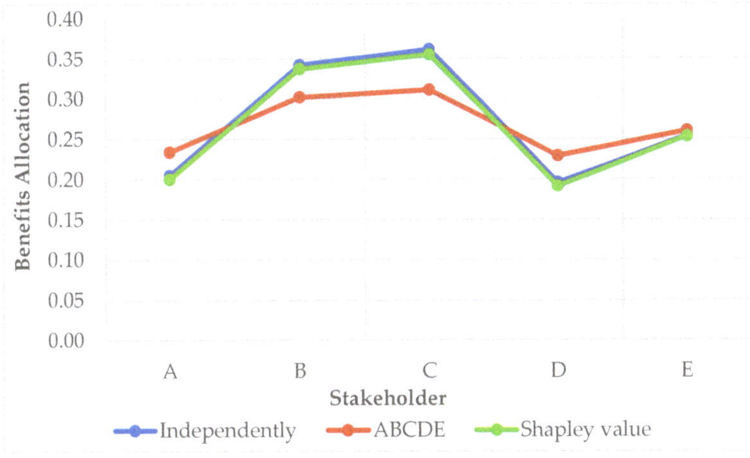

Figure 6. Comparison of coalition distributions.

This is an ideal recommended distribution that would give all stakeholders benefits; otherwise the benefits would only be distributed amongst those who entered a coalition. Some participants (A, D and E) are suggested to decrease their degree of benefit in order to maintain the coalition, since otherwise the other stakeholders might be too unsatisfied with the outcome (their share of the benefits) and believe that their benefits might increase by working on their own (which would not be possible because the entire payoffs model would disintegrate). Thus, some stakeholders are expected to forego a part of their benefits in order that the benefits would be allocated more fairly, while those who contribute more to the coalition can expect to receive higher benefits. This expected increase and decrease of benefits is consistent with previously reported research [60], which show a fair sharing of savings in energy from intercompany heating and cooling integration.

Finally, these results mean that increasing the satisfaction of stakeholders B and C could ensure successful cooperation. To do that it is recommended to trace back those indicators which these stakeholders find more important and work on maximising their performance depending on their objective. For example, by focusing on increasing the GVA impact, despite having little effect on the satisfaction of stakeholder B, it will significantly increase that of stakeholder C. Likewise, reducing further MSW generation increases the satisfaction levels of stakeholder B. It should be noted that these actions do not negatively affect other stakeholders, but will continue contributing to improve their satisfaction levels and thus encourage cooperation towards the NSP scenario [49]. As suggested elsewhere [22], creating equitable benefit and cost distribution to stakeholders in MSWM can increase cooperation and ultimately, the system's sustainability.

5. Conclusions

Even when stakeholders share a common goal, e.g., adopting Circular Economy (CE), conflicting objectives and priorities between different stakeholders are expected to arise. By providing evidence on stable (equilibrium) and optimal decisions, this paper contributes to the decision-making process by proposing a hybrid methodology that attempts to encourage cooperation between stakeholders to adopt CE principles in Municipal Solid Waste Management (MSWM) in cities. This method facilitates the incorporation of all stakeholders' views by considering their multiple and sometimes conflicting priorities. It balances the overall decision-making process by harmonising government technical knowledge, private sector profit-led activities and general public needs.

The efficacy of the proposed framework has been demonstrated with a case study of hypothetically built CE scenarios in Tyseley, Birmingham, UK. The five most influential stakeholder groups were identified and asked to rank nine selected CE indicators that measured the performance of four constructed future scenarios. The subjective and objective weights were calculated for the stakeholders and scenarios, respectively, and these were then used to obtain the stakeholders' preferability indexes and rank their scenarios preferences. The most preferred (or optimal) selection of scenarios was determined using a Nash equilibrium, and the analysis of possible coalitions and the most efficient allocation of benefits was performed using the Shapley value methodology. Thus, the scope of application is to support group decision-making in CE scenarios evaluation, and so it is aimed at the MSWM of cities when multiple stakeholders have different priorities towards future urban scenarios based on CE indicators.

The utilisation of AHP for both the subjective and objective weights not only considers the views and understanding of the stakeholders, but also uses the impartial data of the constructed CE scenarios. The Shapley value allocation of benefits yields a result where all stakeholders share a portion of the benefits; in other words, no coalition where a stakeholder is missing produced an optimal result. However, Cooperative Game Theory (CGT) assumes participants are willing to cooperate and agree on forming coalitions. If stakeholders desert the agreement, the coalition and its benefits model breaks down and jeopardises the possibility of reaching the most preferred, or optimal, scenario.

There are certain factors in decision-making that are extremely difficult to identify and measure; for example, the subjective views on employment in the particular area of Tyseley as briefly presented in the discussions section. In rationality there is no room for human emotions or subjective views. This is a limitation of the proposed method and of Game Theory (GT), as they are both based on the assumptions that actors are intelligent and rational; they have the same information and can make inferences about it, and they will always seek to maximise their utility, respectively. However, in practice, most actors have limited rationality [61] as "rational decision making" and the "rational planning process" assure; meaning that their decisions are bounded by their limited cognitive capacity, restricted time for decision-making and/or by incomplete information [62]. Whilst the proposed method complies with these GT assumptions, it is acknowledged that they also agree with the criticism from rational decision making because it has been widely debated that decision-making is not always rational [63]. However, rational decision making is widely applied in other social and economic disciplines that fall beyond the scope of this study, and that is why it is not explored in more depth.

The study involved the five most influential actors in the particular TEP site; the results were able to come to a most optimal combined scenario for all participants. While it is relatively uncomplicated to define a comprehensive stakeholder directory, it is difficult to predict whether all actors will continue to comply with GT principles (regarding cooperation and willingness to compromise) later in the decision-making process. In this respect, the proposed framework does not consider multiple stages in the decision-making process or the possibility that new stakeholders might be introduced at later stages of the decision-making process. It is also recommended that future research should attempt to measure the awareness of stakeholders towards CE, their willingness to cooperate (accept/pay) to achieve such a CE transition, their level of trust towards other stakeholders, and their perceptions on their counterparts (e.g., more or less powerful and willing or not to forego benefits to bring fairer distributions).

The proposed methodological framework attempts to provide evidence of how the joint selection of the most sustainable scenario could lead to its realisation, and consequently, formulate recommendations to successfully achieve it. It certainly is not the solution to complicated decision-making processes; however, it facilitates them by making the difficult decisions more transparent. In essence, the method represents a single stage of the decision-making process, where it is necessary to converge on a preferred scenario by adjusting the stakeholder satisfaction levels whilst enhancing indicator performance and without damaging the overall decision-making process. However, negotiations might still be fruitless without extensive communication and the development of a common understanding between stakeholders. It is, therefore, recommended that such an investment of prior effort and meticulous preparation through adoption of this methodology is likely to lead to the CE outcomes to which we all already do, or should, aspire.

Author Contributions: Responsible for the Introduction were P.G.P.-A., D.V.L.H., and C.D.F.R.; Methodology, P.G.P.-A., D.V.L.H., and C.D.F.R.; Case Study, P.G.P.-A., D.V.L.H., and C.D.F.R.; Interviews, P.G.P.-A.; Discussion, P.G.P.-A., D.V.L.H., and C.D.F.R.; Conclusions, P.G.P.-A., D.V.L.H., and C.D.F.R.; writing—original draft preparation, P.G.P.-A.; supervision, revision, editing and proofreading, D.V.L.H., and C.D.F.R. All authors have contributed substantially to the work reported. All authors have read and agree to the published version of the manuscript.

Funding: This research received the financial support of the UK EPSRC under grants EP/J017698/1 (Liveable Cities), EP/K012398/1 (iBUILD), and EP/R017727/1 (Coordination Node for UKCRIC). The first author would also like to thank his sponsor, CONACYT-SENER, for funding his doctoral studies at University of Birmingham under scholarship 305047/441645.

Acknowledgments: The authors gratefully acknowledge the participation of the interviewees from the Tyseley Energy Park (TEP), the consultation support from the research team at Smart Villages Lab at University of Melbourne, Australia, and the feedback from various researchers at icRS-Cities 2019, Adelaide, Australia.

Conflicts of Interest: No potential conflict of interest was reported by the authors. The funders had no role in the design of the study; in the collection, analyses, or interpretation of data; in the writing of the manuscript or in the decision to publish the results.

References

1. Palafox-Alcantar, P.G.; Lee, S.E.; Bouch, C.; Hunt, D.V.L.; Rogers, C.D.F. *The Little Book of Circular Economy in Cities*; Imagination Lancaster: Lancaster, UK, 2017; ISBN 978-0-70442-950-5.
2. Palafox-Alcantar, P.G.; Hunt, D.V.L.; Rogers, C.D.F. The complementary use of game theory for the circular economy: A review of waste management decision-making methods in civil engineering. *Waste Manag.* **2020**, *102*, 598–612. [CrossRef] [PubMed]
3. Rogers, C.D.F. Engineering Future Liveable, Resilient, Sustainable Cities Using Foresight. *Proc. Inst. Civ. Eng. Eng.* **2018**, *171*, 3–9. [CrossRef]
4. Circle Economy. *The Circularity Gap Report*; European Union: Amsterdam, Netherlands, 2018.
5. Kirchherr, J.; Piscicelli, L.; Bour, R.; Kostense-Smit, E.; Muller, J.; Huibrechtse-Truijens, A.; Hekkert, M. Barriers to the Circular Economy: Evidence from the European Union (EU). *Ecol. Econ.* **2018**, *150*, 264–272. [CrossRef]
6. Veleva, V.; Bodkin, G. Corporate-entrepreneur collaborations to advance a circular economy. *J. Clean. Prod.* **2018**, *188*, 20–37. [CrossRef]

7. Lee, D.H. Econometric assessment of bioenergy development. *Int. J. Hydrogen Energy* **2017**, *42*, 27701–27717. [CrossRef]
8. Kastner, C.A.; Lau, R.; Kraft, M. Quantitative tools for cultivating symbiosis in industrial parks: A literature review. *Appl. Energy* **2015**, *155*, 599–612. [CrossRef]
9. Soltani, A.; Sadiq, R.; Hewage, K. Selecting sustainable waste-to-energy technologies for municipal solid waste treatment: A game theory approach for group decision-making. *J. Clean. Prod.* **2016**, *113*, 388–399. [CrossRef]
10. Luo, M.; Song, X.; Hu, S.; Chen, D. Towards the sustainable development of waste household appliance recovery systems in China: An agent-based modeling approach. *J. Clean. Prod.* **2019**, *220*, 431–444. [CrossRef]
11. Friedrich, E.; Trois, C. Current and future greenhouse gas (GHG) emissions from the management of municipal solid waste in the eThekwini Municipality—South Africa. *J. Clean. Prod.* **2016**, *112*, 4071–4083. [CrossRef]
12. Ripa, M.; Fiorentino, G.; Vacca, V.; Ulgiati, S. The relevance of site-specific data in Life Cycle Assessment (LCA). The case of the municipal solid waste management in the metropolitan city of Naples (Italy). *J. Clean. Prod.* **2017**, *142*, 445–460. [CrossRef]
13. Fei, F.; Wen, Z.; Huang, S.; De Clercq, D. Mechanical biological treatment of municipal solid waste: Energy efficiency, environmental impact and economic feasibility analysis. *J. Clean. Prod.* **2018**, *178*, 731–739. [CrossRef]
14. Estay-Ossandon, C.; Mena-Nieto, A.; Harsch, N. Using a fuzzy TOPSIS-based scenario analysis to improve municipal solid waste planning and forecasting: A case study of Canary archipelago (1999–2030). *J. Clean. Prod.* **2018**, *176*, 1198–1212. [CrossRef]
15. De Feo, G.; De Gisi, S. Using an innovative criteria weighting tool for stakeholders involvement to rank MSW facility sites with the AHP. *Waste Manag.* **2010**, *30*, 2370–2382. [CrossRef] [PubMed]
16. Soltani, A.; Hewage, K.; Reza, B.; Sadiq, R. Multiple stakeholders in multi-criteria decision-making in the context of Municipal Solid Waste Management: A review. *Waste Manag.* **2015**, *35*, 318–328. [CrossRef] [PubMed]
17. Li, N.; Zhao, H. Performance evaluation of eco-industrial thermal power plants by using fuzzy GRA-VIKOR and combination weighting techniques. *J. Clean. Prod.* **2016**, *135*, 169–183. [CrossRef]
18. Sabaghi, M.; Mascle, C.; Baptiste, P. Evaluation of products at design phase for an efficient disassembly at end-of-life. *J. Clean. Prod.* **2016**, *116*, 177–186. [CrossRef]
19. Kaźmierczak, U.; Blachowski, J.; Górniak-Zimroz, J. Multi-Criteria Analysis of Potential Applications of Waste from Rock Minerals Mining. *Appl. Sci.* **2019**, *9*, 441. [CrossRef]
20. Strantzali, E.; Aravossis, K.; Livanos, G.A.; Nikoloudis, C. A decision support approach for evaluating liquefied natural gas supply options: Implementation on Greek case study. *J. Clean. Prod.* **2019**, *222*, 414–423. [CrossRef]
21. Grimes-Casey, H.G.; Seager, T.P.; Theis, T.L.; Powers, S.E. A game theory framework for cooperative management of refillable and disposable bottle lifecycles. *J. Clean. Prod.* **2007**, *15*, 1618–1627. [CrossRef]
22. Karmperis, A.C.; Aravossis, K.; Tatsiopoulos, I.P.; Sotirchos, A. Decision support models for solid waste management: Review and game-theoretic approaches. *Waste Manag.* **2013**, *33*, 1290–1301. [CrossRef]
23. Chen, F.; Chen, H.; Guo, D.; Han, S.; Long, R. How to achieve a cooperative mechanism of MSW source separation among individuals—An analysis based on evolutionary game theory. *J. Clean. Prod.* **2018**, *195*, 521–531. [CrossRef]
24. Rogers, C.D.F.; Hunt, D.V.L. Realising visions for future cities: An aspirational futures methodology. *Proc. Inst. Civ. Eng. Urban Des. Plan.* **2019**, *172*, 125–140. [CrossRef]
25. UKCRIC. *Evidence Base and Portfolio of Tools and Techniques Underpinning UKCRIC's Research Capability in Cities*; Collaboratorium for Research on Infrastructure and Cities: London, UK, 2019; Available online: www.ukcric.com (accessed on 27 March 2020).
26. Bocken, N.; de Pauw, I.; Bakker, C.; van der Grinten, B. Product design and business model strategies for a circular economy. *J. Ind. Prod. Eng.* **2016**, *33*, 308–320. [CrossRef]
27. Rizos, V.; Behrens, A.; van der Gaast, W.; Hofman, E.; Ioannou, A.; Kafyeke, T.; Flamos, A.; Rinaldi, R.; Papadelis, S.; Hirschnitz-Garbers, M.; et al. Implementation of Circular Economy Business Models by Small and Medium-Sized Enterprises (SMEs): Barriers and Enablers. *Sustainability* **2016**, *8*, 1212. [CrossRef]

28. Witjes, S.; Lozano, R. Towards a more Circular Economy: Proposing a framework linking sustainable public procurement and sustainable business models. *Resour. Conserv. Recycl.* **2016**, *112*, 37–44. [CrossRef]
29. Weiner, E.; Brown, A. Stakeholder Analysis for Effective Issues Management. *Plan. Rev.* **1986**, *14*, 27–31. [CrossRef]
30. Macharis, C.; Turcksin, L.; Lebeau, K. Multi actor multi criteria analysis (MAMCA) as a tool to support sustainable decisions: State of use. *Decis. Support Syst.* **2012**, *54*, 610–620. [CrossRef]
31. Hunt, D.V.L.; Lombardi, D.R.; Rogers, C.D.F.; Jefferson, I. Application of sustainability indicators in decision-making processes for urban regeneration projects. *Proc. Inst. Civ. Eng. Eng. Sustain.* **2008**, *161*, 77–91. [CrossRef]
32. Valenzuela-Venegas, G.; Salgado, J.C.; Díaz-Alvarado, F.A. Sustainability indicators for the assessment of eco-industrial parks: Classification and criteria for selection. *J. Clean. Prod.* **2016**, *133*, 99–116. [CrossRef]
33. Saidani, M.; Yannou, B.; Leroy, Y.; Cluzel, F.; Kendall, A. A taxonomy of circular economy indicators. *J. Clean. Prod.* **2019**, *207*, 542–559. [CrossRef]
34. Leach, J.M.; Braithwaite, P.A.; Lee, S.E.; Bouch, C.J.; Hunt, D.V.L.; Rogers, C.D.F. Measuring urban sustainability and liveability performance: The City Analysis Methodology. *Int. J. Complex. Appl. Sci. Technol.* **2016**, *1*, 86. [CrossRef]
35. Lombardi, D.; Leach, J.; Rogers, C.; Aston, R.; Barber, A.; Boyko, C.; Brown, J.; Bryson, J.; Butler, D.; Caputo, S.; et al. *Designing Resilient Cities: A Guide to Good Practice*; IHS BRE Press: Bracknell, UK, 2012; ISBN 9781848062535.
36. Hunt, D.V.L.; Lombardi, D.R.; Atkinson, S.; Barber, A.R.G.; Barnes, M.; Boyko, C.T.; Brown, J.; Bryson, J.; Butler, D.; Caputo, S.; et al. Scenario Archetypes: Converging Rather than Diverging Themes. *Sustainability* **2012**, *4*, 740–772. [CrossRef]
37. Hunt, D.V.L.; Jefferson, I.; Rogers, C.D.F. Assessing the Sustainability of Underground Space Usage—A Toolkit for Testing Possible Urban Futures. *J. Mt. Sci.* **2011**, *8*, 211–222. [CrossRef]
38. Saaty, T.L. *The Analytic Hierarchy Process*; McGraw-Hill: New York, NY, USA, 1980.
39. Feyzi, S.; Khanmohammadi, M.; Abedinzadeh, N.; Aalipour, M. Multi- criteria decision analysis FANP based on GIS for siting municipal solid waste incineration power plant in the north of Iran. *Sustain. Cities Soc.* **2019**, 47. [CrossRef]
40. Doloi, H. Application of AHP in improving construction productivity from a management perspective. *Constr. Manag. Econ.* **2008**, *26*, 839–852. [CrossRef]
41. Saaty, T.L.; Tran, L.T. On the invalidity of fuzzifying numerical judgments in the Analytic Hierarchy Process. *Math. Comput. Model.* **2007**, *46*, 962–975. [CrossRef]
42. Nash, J. Non-Cooperative Games. *Ann. Math.* **1951**, *54*, 286–295. [CrossRef]
43. Madani, K.; Hipel, K.W. Non-Cooperative Stability Definitions for Strategic Analysis of Generic Water Resources Conflicts. *Water Resour. Manag.* **2011**, *25*, 1949–1977. [CrossRef]
44. Porter, R.; Nudelman, E.; Shoham, Y. Simple search methods for finding a Nash equilibrium. *Games Econ. Behav.* **2008**, *63*, 642–662. [CrossRef]
45. McKelvey, R.; McLennan, A. No Computation of equilibria in finite games. *Handb. Comput. Econ.* **1996**, *1*, 87–142. [CrossRef]
46. McKelvey, R. *A Liapunov Function for Hash Equilibria*; California Institute of Technology: Pasadena, CA, USA, 1998.
47. Govindan, S.; Wilson, R. A global Newton method to compute Nash equilibria. *J. Econ. Theory* **2003**, *110*, 65–86. [CrossRef]
48. Govindan, S.; Wilson, R. Computing Nash equilibria by iterated polymatrix approximation. *J. Econ. Dyn. Control* **2004**, *28*, 1229–1241. [CrossRef]
49. Asgari, S.; Afshar, A.; Madani, K. Cooperative Game Theoretic Framework for Joint Resource Management in Construction. *J. Constr. Eng. Manag.* **2014**, 140. [CrossRef]
50. Shapley, L.S. A value for n-Person games. In *Contributions to the Theory of Games II*; Kuhn, H.W., Tucker, A.W., Eds.; Princeton University Press: Princeton, NJ, USA, 1953; pp. 307–317.
51. Cano-Berlanga, S.; Giménez-Gómez, J.M.; Vilella, C. Enjoying cooperative games: The R package GameTheory. *Appl. Math. Comput.* **2017**, *305*, 381–393. [CrossRef]
52. Lee, S.E.; Quinn, A.D.; Rogers, C.D.F. Advancing City Sustainability via its Systems of Flows: The Urban Metabolism of Birmingham and its Hinterland. *Sustainability* **2016**, *8*, 220. [CrossRef]

53. Wang, J.J.; Jing, Y.Y.; Zhang, C.F.; Zhao, J.H. Review on multi-criteria decision analysis aid in sustainable energy decision-making. *Renew. Sustain. Energy Rev.* **2009**, *13*, 2263–2278. [CrossRef]
54. Behera, S.K.; Kim, J.H.; Lee, S.Y.; Suh, S.; Park, H.S. Evolution of "designed" industrial symbiosis networks in the Ulsan Eco-industrial Park: "research and development into business" as the enabling framework. *J. Clean. Prod.* **2012**, *29–30*, 103–112. [CrossRef]
55. Birmingham City Council. *From Waste to Resource: A Sustainable Strategy for 2019*; A Report from Overview & Scrutiny: Birmingham, UK, 2014.
56. International Synergies. *Industrial Symbiosis Study for the Tyseley Environmental Enterprise Zone*; International Synergies Ltd.: Birmingham, UK, 2013.
57. Goepel, K.D. Implementing the Analytic Hierarchy Process as a Standard Method for Multi-Criteria Decision Making in Corporate Enterprises—A New AHP Excel Template with Multiple Inputs. In Proceedings of the International Symposium on the Analytic Hierarchy Process, Kuala Lumpur, Malaysia, 23–26 June 2013.
58. Rogers, C.D.F. *Chapter 11: Cities*; Government Office for Science: London, UK, 2017.
59. Gouldson, A.; Kerr, N.; Kuylenstierna, J.; Topi, C.; Dawkins, E.; Pearce, R. *The Economics of Low Carbon Cities: A Mini-Stern Review for Birmingham and the Wider Urban Area*; Centre for Low Carbon Futures: Leeds, UK, 2013.
60. Hiete, M.; Ludwig, J.; Schultmann, F. Intercompany Energy Integration: Adaptation of Thermal Pinch Analysis and Allocation of Savings. *J. Ind. Ecol.* **2012**, *16*, 689–698. [CrossRef]
61. Li, J.S.; Fan, H.M. Symbiotic Cooperative Game Analysis on Manufacturers in Recycling of Industrial Wastes. *Appl. Mech. Mater.* **2013**, *295–298*, 1784–1788. [CrossRef]
62. Lee, C. Bounded rationality and the emergence of simplicity amidst complexity. *J. Econ. Surv.* **2011**, *25*, 507–526. [CrossRef]
63. Simon, H.A. Game Theory and the Concept of Rationality. In *Herbert Simon Collection*; Carnegie Mellon University: Pittsburgh, PA, USA, 1999.

Article

Exploring Citizens' Actions in Mitigating Climate Change and Moving toward Urban Circular Economy. A Multilevel Approach

Adriana AnaMaria Davidescu [1,2,*], Simona-Andreea Apostu [1,3] and Andreea Paul [4]

[1] Department of Statistics and Econometrics, Bucharest University of Economic Studies, Romana Square, 15–17 Dorobanți St., Sector 1, 010552 Bucharest, Romania; simona.apostu@csie.ase.ro

[2] Labour Market Policies Department, National Scientific Research Institute for Labour and Social Protection, 6–8, Povernei Street, 010643 Bucharest, Romania

[3] Institute of National Economy, Romanian Academy House, 13 September 13, 050711 Bucharest, Romania

[4] Department of International Economic Relations, Bucharest University of Economic Studies, 15–17 Dorobanti St., Sector 1, 010552 Bucharest, Romania; andreea.paul@inaco.ro

* Correspondence: adrianaalexandru@yahoo.com

Received: 22 July 2020; Accepted: 3 September 2020; Published: 11 September 2020

Abstract: Urbanization and climate change are requiring cities to find novel pathways to a sustainable future, and therefore the urban context may accelerate the conversion to a circular economy. In this sense, climate change is a considerable threat to the environment, affecting both human and natural systems, and in this context individuals have a very important role. Therefore, the paper aims to investigate, on the one hand, what determines people to undertake specific actions in fighting climate change and, on the other hand, what determines some people to engage in adopting multiple actions exhibiting extra mitigation behaviour compared to others, paving the way to an urban circular economy. In order to do that, multilevel logistic regression analysis using hierarchical data (individuals grouped in counties), reflecting group variability and group-level characteristics effects on outcomes at individual level has been applied. Special attention was given to modernisation thesis validation, stipulating that citizens from more developed and modernized countries are expected to manifest a higher level of extra mitigation compared to inhabitants of less-modernized nations. The empirical results revealed the positive association of pro-environmental factors, socio-demographic and economic factors with both specific and extra mitigation behaviour in fighting climate change. An important finding of the empirical research highlighted the validation of the modernisation thesis, even if partially, and the reinforcement of the modernisation thesis impact on the extra mitigation behaviour determined by the urban area segmentation. The extra commitment behaviour reflected by citizens' multiple actions in fighting climate change ensures progress to a circular economy through its contribution to waste reduction, eco-shopping increase, on eco-friendly transportation increase or domestic energy reduction. We believe that a shift in citizens' attitude towards climate change is needed, taking into account that a lot must be done" to effectively respond to climate change, paving the way for the circular economy.

Keywords: circular economy; urbanization and climate change; multilevel logistic regression; citizens; extra mitigation behavior; EU member states

1. Introduction

In the context of urbanization and excessive pollution, on the path to achieving sustainable development, the principles of circular economy become increasingly important and also necessary. The circular economy implies material reducing, reusing, remanufacturing and recycling, but the transition is not possible without including environmental considerations, such as climate change. Potential synergies between circular material use, climate change mitigation and the halting of biodiversity loss are increasingly recognized [1].

Urbanization and climate change are determining cities to find novel pathways in order to lead to a sustainable future, therefore the urban context may precipitate the conversion to a circular economy. In this sense, climate change is a considerable danger to the environment, affecting both human and natural systems [2], and in this context individuals have a very important role [3].

Climate change needs to be mitigated, requiring a change in personal behavior, from levels of action or limited action to broader and higher levels of behavioural involvement [4]. A further way of tackling climate change is represented by the adoption of auxiliary actions leading to a higher level of engagement in climate change mitigation going beyond what most people do [5].

Cities represent the link to circular development, the city administrations being able to specify and inform a circular vision and strategy. Cities can provide opportunities for accessing, purchasing, promoting and stimulating circular solutions.

The movement to a circular economy is not easy, particularly for companies in which all the operations are deeply subordinated to the linear approach, production processes must first be transformed from linear to circular [6] and, accordingly, the implications of such a transformation refers to changes of processes, additional investments, modifications in equipment and production, shifts in raw materials, as well as personnel re-training with consequences on extensive value chain.

The current model of wasting resources involves depleting the natural capital of the Earth, generating irrevocable and alarming effects on both environment and climate.

Climate change is a serious environmental challenge, the reduction of greenhouse gas (GHG) emissions representing an important target that can be acquired only by the involvement in actions targeting the increase of resource efficiency, prolonging the lifecycle of buildings and goods, enhancement of recycling and reuse together with a decline in primary raw materials [1].

Therefore, the main objective of the research is two-fold: on the one hand to enhance understanding of the main environmental, socio-demographic and economic determinants of citizens' specific actions to mitigate climate change and, on the other hand, to investigate factors related to a higher level of extra commitment in fighting climate change in an urban context.

Thus, the paper aims to answer to the following main question: "What determines some people engaging in adopting multiple actions exhibiting extra mitigation behaviour compared to others?" Finding a reliable respond to this question will also provide solutions that will help stakeholders to make the transition to an urban circular economy.

More in depth, the research aims also to respond to specific research questions through the results provided by the multilevel approach: *what are the main characteristics of pro-environmental behaviour associated to an increased level of engagement in adopting specific actions in fighting climate change as well as an extra mitigation behaviour? What are the main socio-demographic characteristics associated with a higher level of engagement in adopting specific actions in fighting climate change as well as an extra mitigation behaviour? What are the main economic characteristics associated with an increased level of engagement in adopting specific actions in fighting climate change as well as an extra mitigation behaviour? Is extra mitigation behaviour to climate change higher in more modernized developed economies? What is the extent of between-country variation in adopting taking specific actions as well as in developing extra mitigation behaviour in fighting climate change? Can between-country differences in fighting climate change be explained by differences in individual characteristics? Does the effect of modernization thesis on the extra mitigation behavior depends on the segmentation of large town* vs. *small or medium town?*

In order to mitigate climate change and also to pave the way to circular economy, the role of citizens is fundamental and a shift in personal behaviour is required. In combating climate change, the actions need to cover multiple measures and to target the implications of the main actors—local governments, business sector, civic society, environmental organisations and European institutions.

In the paper, the focus is on the role of citizens, the personal mitigation behaviour referring to voluntary behavioural responses to climate change.

The article makes its contribution to the literature in the following four ways. Firstly, it investigates factors related to specific actions made by citizens to combat climate change, revealing the main individual determinants that could lead citizens to make a certain action by the analysis of 13 self-reported actions, revealing also how the actions taken to combat climate change could support the shift to a circular economy.

Secondly, the study investigates factors related to citizens' profile characteristics leading them to adopt specific actions to address climate change, highlighting the differences among individual actions.

Thirdly, the study explores the individual extra mitigation behaviour in fighting climate change, analysing the individual behaviour in terms of cumulative actions, taking into account that according to Ortega et al. [5], a lot must be done to effectively address climate change. More in depth, the research examines the between-country variations related to both citizens' actions in tackling climate change, as well as an extra engagement level in combating climate change using a staged multi-level logistic regression model based on hierarchical data (individuals within countries). The utility of this category of models have been acknowledged also by Ortega-Egea [5] who mentioned that multilevel models are the most appropriate in tackling environmental researches.

In particular, the research makes a contribution to climate change research by testing the modernization thesis according to which an extra mitigation behavior for climate change is higher in more modernized developed economies, highlighting also how a broader and greater level of commitment can pave the way to a circular economy.

Fourthly, for exploring country discrepancies in both specific and extra mitigation behaviour to fight climate change, the dataset of the recent Eurobarometer 91.3 Rule of law and climate change has been used [7], having the advantage of using the most recent large-scale survey conducted at the end of 2019. In such a way, the study contributes to the research field through an advance of current understanding of climate change extra mitigation behaviour reinforcing also the main implications for supporting the shift to a circular economy. The main results of the research represent a useful tool mainly for policy-makers that need to focus on expanding the engagement level in tackling climate change both among citizens and across countries.

The paper is organized as follows. The Section entitled "Literature review" offers an overview of the most proper studies regarding climate change mitigation and its implications for the transition to a circular economy. Section 3 offers additional information about the data used in the analysis providing also a brief description of the methodology used within the paper. Section 4 is divided into five main sub-sections presents the descriptive profile of the respondents, a brief analysis of climate change mitigation actions among European Union (EU) countries and also the sub-sections of multilevel models results, summary of findings in relation to specific mitigation and extra mitigation behaviour and discussions in relation to the man findings. The paper ends with concluding remarks and also by highlighting limitations and recommendations for further study.

2. Literature Review

The circular economy emerged as a response to maximize the reuse of assets and products and minimize their depreciation [8] becoming the driving force of sustainable land development, aiming to maximize resource efficiency together with impact reduction on environment [9].

The circular economy put together four main actions: reduce, reuse and recycle and remanufacture of materials in the processes of production, distribution and consumption [10,11]. The circular economy relies mainly on resource and environmental efficiency, aiming to provide key measures that will assure the transition to a greener and more sustainable economy [12].

Korhonen et al. [13] define the circular economy as an approach aiming to fight against environmental challenges and encourage sustainable development, reducing both the input of virgin materials and waste production by closing the economic and ecological curves of resource courses [14].

The circular economy achieve optimal production by simultaneously minimizing the use of natural resources, minimizing waste by reusing production, and minimizing pollution by recycling and restoring industrial unnecessary waste [15].

The circular path is more than just a recycling and environmental issue, it represents a new approach to thinking about how to grow and develop without resorting only to spending resources [16], but reusing them. Reuse supports resource efficiency, reducing air, water and soil pollution during the lifecycle of the product [17]. In this context, circular economy can play the fundamental role of driving force, strongly supporting the attainment of Sustainable Development Goals [18].

There are many definitions for climate change in the literature, but the most prominent has been provided by the Ellen MacArthur Foundation [19], according to which circular economy represents an industrial system designed and intended to be restorative or regenerative, replacing restoration with transition to renewable energy usage. This excludes the use of toxic chemicals, which impair reuse, and eliminate waste [20,21].

The circular economy represents a model of generating and waste, involving the distribution, rental, reusing, repair, reconditioning and recycling of materials and products as long as possible, aiming also to increase the product lifecycle.

Therefore, the circular economy has 3 dimensions:

sustainable food system that can lead to improved efficiency, solving logistical problems and protecting food security in the agricultural sector;
resource management and waste optimization that can improve the waste-recycling process;
reducing pollution that has raised awareness about polluted cities and the associated health risks [22].

The circular economy relies on three major principles: waste and pollution reduction, preserve using products and materials; reestablishing natural systems [19].

The transition to a circular economy implies targeted changes implying all three actors: civic society, governments and business sector, with consequences on different subsystems: energy, logistics and financial subsystems and clear guidance and monitoring [23]. The circular economy involves both public actors responsible for sustainable development and planning, as well as businesses seeking economic, social and environmental results and society that should ask itself about its real needs [24].

The urban population is constantly growing, so cities must meet the criteria of sustainability. UN-HABITAT (2011) [25] reported that urban areas generate 70% of global greenhouse gas emissions and 50% of global waste [26]. In this sense, measures are being taken in the field of sustainability, cities and communities being encouraged to apply ingenious strategies to address circularity on urban scale [27], reducing the breadth of global waste streams and emissions [28].

The model of circular economy implies reusing, reducing, recycling and remanufacturing [29] (Figure 1), representing sustainable ecological growth by extending the life of products and materials and reducing waste. The path to circular economy should also include environmental considerations, such as climate change [30]. The circular model has many environment, climate, social and economic benefits, one being the mitigation of climate change [1].

In 2019, the World Economic Forum stated that the circular economy could represent an important tool to avoid climate change [31]. Reducing greenhouse gas emissions in the circular economy is reflected by increasing resource efficiency, by raising the life of construction and goods, increasing recycle and reuse, and a complete reduction in using primary raw materials. According to the Committee on Climate Change (CCC), the transition to a circular economy must be a part of climate change. As part of Covid-19 recovery plan that prioritizes the climate crisis, the CCC has urged the governments to hurry the transition to a circular economy, reiterating its call for a ban on biodegradable wastes to landfill by 2025 [32].

A major global challenge is represented by climate change [33], requiring the international community to adopt mitigation strategies having as their main target limiting the average increase of global temperature [34,35]. Climate change is a major, anthropically caused environmental risk, with important effects for human and natural systems [36].

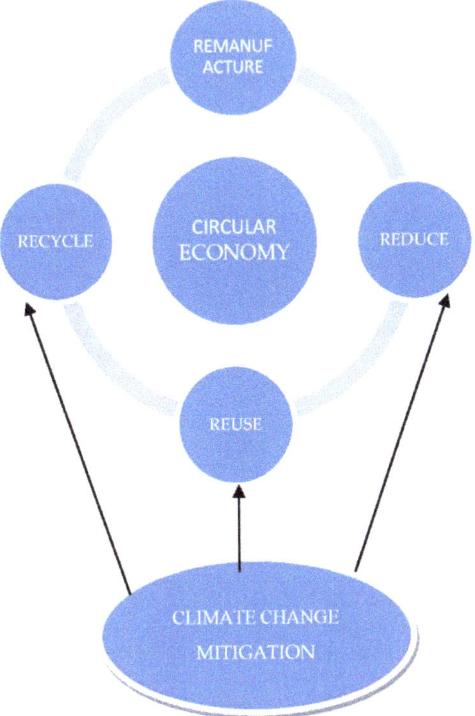

Figure 1. The implications of climate change mitigation on the transition to circular economy. Source Author's contribution.

Under the Framework Convention on Climate Change, climate change is defined as a climate modification attributed to human activity, altering the the global atmosphere composition [37]. According to the Intergovernmental Panel on Climate Change, climate change is defined as any modification regarding climate caused by natural variability or human activity [38].

Climate change is significantly influenced by humans, the main changes occurring in the atmospheric composition. These are caused by emissions associated with energy consumption, urbanism and land-use changes, being very important at local and regional level. Although certain progress have been achieved in the processes of monitoring and exploring climate change, several scientific, technical and institutional obstacles still have to be overcome in order to design, plan for, adapt to and mitigate the impact of climate change [39].

Climate change mitigation calls for cooperation both national governments and consumers, the mitigation effort being concentrated on reducing sources and amplifying greenhouse gas sinks, requiring concerted efforts of both national and international institutions that take quite a long time [40]. One potential solution for urgent measures is the increase in the level of citizens' commitment by adopting voluntary more-sustainable and low-carbon lifestyle alternatives.

Through a higher level of individuals' engagement, a higher level of receptivity and response to change messages, emissions from home using up and personal transportation could be significantly attenuated [41].

The most important factor causing climate change is excessive energy consumption, so in order to adopt a new low-carbon paradigm, energy reduction is needed, both on businesses and civil societies [3]. Even if climate change is generally seen as a global issue and municipal governments have their local agenda as an important tool to control greenhouse gas emissions, the mitigation behaviour cannot be improved without the help of state and national policy changes [42].

In this fight against climate change, the contribution of citizens becomes increasingly important [43], they acquiring the role of agents of change in field of environment [44].

Although the anthropogenic cause of climate change is scientifically proven, people consider different causes regarding climate change, such as: natural processes, human activities, determining both risk perception and mitigation behaviour [45].

In industrialized countries, the principal cause of climate change is the lifestyle [41], with high level of carbon emissions, with households responsible for 20% of total greenhouse gas emissions [46] and over a quarter of the final energy consumed [47]. Therefore, individuals have a very important role in the face of climate change [3], either adapting to potential and unavoidable climate impacts, or decreasing GHG emissions to avoid further damage [48].

The literature is focused on different specific actions undertaken to combat climate change in relation to the transition to a circular economy, grouped in four main categories: eco-friendly transportation referring to: low fuel consumption car, electric car, alternatives to car use, travel carbon footprint considered; domestic energy reduction referring to: insulated home better, low-energy home, energy efficiency households (EFF HH) appliances, switched energy supplier, energy-saving equipment home, installed solar panels home; eco-shopping referring to: food carbon footprint considered; and waste reduction refering to: reduce and separate waste, use less disposable items (Figure 2).

Previous studies targeted very intensively recycling [17,18], reducing the use of the car and adopting another means of transport that supports the environment, [22,49] or explaining involvement in environmental citizenship [19] by various individual factors [12,23–30,50]. The relevance of pro-environmental behavior has proved its stability over time [11,33] being related to ecological behavior [34]. Ortega-Egea et al. [5] analysed the main factors related to an extra mitigation behaviour among individuals, revealing the relevance of gender, age, attitudes, knowledge and political ideology in combating climate change.

Although most people consider climate change and sustainability to be important issues, too few citizens are taking action to stop the growing flow of greenhouse gases and other related issues. This is due to adverse climate infrastructure, and psychological barriers such as: limited knowledge of the issue, ideological conceptions of the world that tend to deter a pro-environmental habit, comparisons with other key people, low costs and time of behavior, discrediting of experts and authorities, perceived risks of change [51].

The models tested in the research propose that environmental factors, socio-demographics and modernisation theory-related factors enhance understanding of citizens' specific actions to mitigate climate change as well as the extra mitigation behavior in fighting climate change in urban context.

The literature review is structured on the four main factors related to climate change mitigation: environmental factors, demographics factors, socio-economic factors and state-level factors.

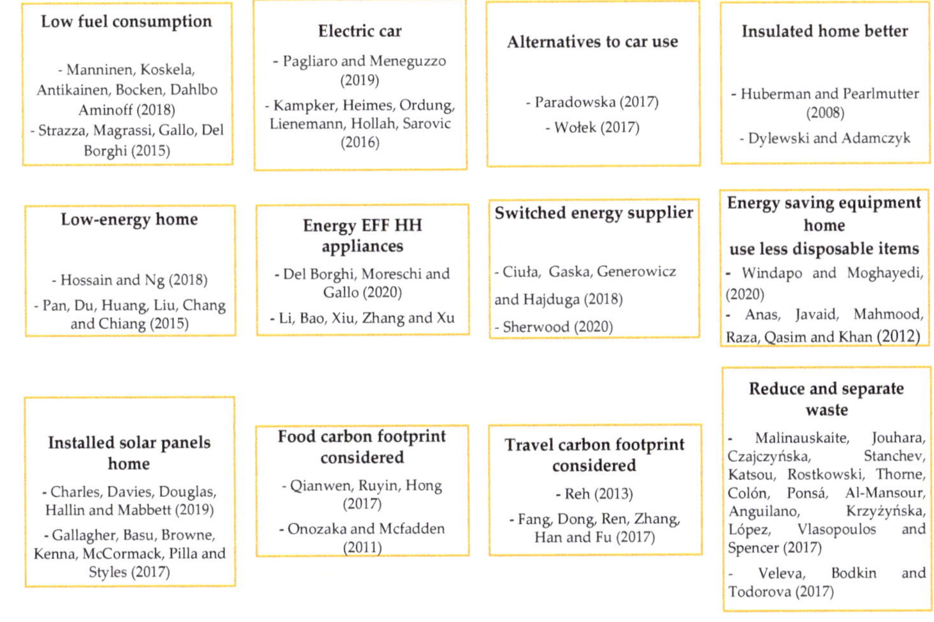

Figure 2. An overview of the most relevant studies concerning specific mitigation behaviour and its relation with the circular economy. Source: Author's contribution.

2.1. Perceived Importance of Climate Change

The pro-environmental and climate change motivated behavior has been analyzed in the literature [52–63]. The concept of global environmentalism [63–65] emerged as a consequence of environmental degradation, joint with pro-environmental actions in wealthy and developing countries [66,67]. The globalization hypothesis was debated in specialized studies [67,68], with some studies [51,54,57–59,68,69] supporting the country significant differences regarding public environmental concern and protection, including climate change mitigation efforts [62].

People do not take any action to mitigate climate change because they do not believe that climate is changing and, if it does, they deny the involvement of human activity [70–73]. According to Tranter and Booth [74], the proportion of climate sceptics was around one-fifth of the interviewed respondents from the dataset of the 2010 International Social Survey Programme, while according to the World Values Survey, the proportion of sceptics is even lower, the general perception being as "very serious" issue [75,76]. The results of 2017 Pew Global Attitudes and Trends survey [77] revealed that almost 61% of respondents from 38 countries acknowledged that climate change is a "major threat" in case of their country and only 23% of them consider climate change a "minor threat".

People's opinion about climate change is influenced by the position of scientists on this issue, the disagreement of scientists on global warming leads to the new indivisibility to global warming and can be reduced for climate policy [78].

Some people do not believe in climate change, in its anthropogenic character and its importance to humans and nature [67,68,74,79,80]. They are climate skeptics and represent at most one-fifth of respondents to the 2010 International Social Survey Program [74], being more prevalent in Anglo-American countries, especially in the United States [81,82]; however, they represent a declining minority [83–85]. Therefore, most people consider climate change to be anthropogenic, real, and costly [86].

2.2. Climate Change Perceived Responsibility

In the context of climate change, responsibility is the obligation of people to achieve their goal of protecting the environment [87]. Regarding how individuals respond to climate change, four typologies can be identified—community, systemic, skeptical and economist; most of them claiming that individuals act based on their perceived responsibility regarding climate change. It is worth mentioning that under these circumstances, behavioural change is seen as an optimal combination of living standards, knowledge of the causes and climate change contributions and harmful emissions intensity, involving actions of individuals as voters and consumers [88].

Responsibility in the case of climate change is also attributed to state institutions, governments and churches taking action so as to reduce dependence on carbon-based energy sources [89,90].

2.3. The Perceived Importance of Climate Actions

The perceived importance in the case of climate change is influenced by environmental position and climate change knowledge, the relationship being mediated by risk perception [91].

Bockarjova and Steg [92] highlighted that measures to combat climate change are associated with health, including quitting smoking or promoting exercise, and changing behavior patterns. In this context, Bergantino and Catalano [93] consider that age, sex, employment status and the number of young children significantly influence the psychological profiles of respondents.

For civil societies fighting against climate change it is very important that citizens are well informed and know about this problem [94]. In this regard, informational efforts have been made to encourage public and voluntary involvement in climate change mitigation actions [95,96]. Other measures that have led to public awareness and preoccupation about climate change are economic incentives, subsidies and related interventions [97].

2.3.1. Gender

Pro-environmental behavior is influenced by gender, with women being more likely to take action to combat climate change [98,99]. Women are more likely to work for the environment in response to climate change by reducing waste and saving energy during the day [55,100]. Other studies have highlighted the greater concern and involvement of women about the environment compared to men [98,101,102]. Ortega-Egea [5] found evidence although weakly that female gender was more inclined to develop an extra mitigation behavior to address climate change.

In some areas, climate policies are insensitive to gender issues and do not take into account the fact that some adaptation actions may be less available for certain groups of women, such as the head of the household [103]. Countries whose development policies aim at the stability of the system by strengthening the status quo, neglects the differentiated vulnerability, adaptability, existing cultural, institutional and political factors of inequality [104].

2.3.2. Age

Environmental behavior is most influenced by age, with young people being those most concerned about climate change [105]. Most studies have found that age is inversely correlated with attitudes and concerns about the environment, with young people being more concerned about climate change [3,106,107], perceiving the seriousness of climate change more acutely [108]. Other studies have highlighted the negative, positive or insignificant link between age and environmental behaviour [48,98]. An important result has been obtained by the study of Ortega-Egea [5] indicating an inverted U-shaped relationship between age and individual level of engagement, citizens aged between 45–56 years old tend to be more inclined to exhibit an extra mitigation by way of acting.

2.3.3. Education

Women, people with a lower level of education, middle-aged, with higher household incomes are more skeptical about the existence of climate change [109]. Many studies have shown a direct correlation between the level of education and environmental behavior [5,98,110], individuals with a high level of education are informed, concerned and involved in pro-environmental activities [98,107,111,112], undertaking actions to mitigate climate change [2,3,41]. Ortega-Egea [5] proved that there is a appositive association between education and extra mitigation behavior.

2.3.4. Marital Status (Children)

Marital status is an important characteristic, particularly in the case of women, that affects the social relationships [113]. Gender intersects with marital status, being discursively produced [109,114] and manifested in the concrete actions of women and men [115].

Chant [116] argues that women from households formed exclusively by women face poverty and vulnerability differently to women from mixed households formed by women and men. While women in households formed exclusively by women often have to deal with the problem of limited resources, women from mixed households have less access and control over household assets.

Englert [117] highlighted that unmarried, divorced and widowed women have more freedom to buy land or take action, marital status having an important role, and in the case of married women their actions depend on the nature of their relationship with the husband. The marital status significantly influences climate change initiatives, mainly for women, and this results has been supported by the study of KatrienVan and Holvoet [103].

2.3.5. Occupation

Confidence in the environment, awareness of the main implications of climate change, high income, employment and old age have led to support for climate change policy [112]. Employment status has no effect on environmental concern [118], a higher remuneration is associated with an increase in the level of pro-environmental engagement [119].

2.3.6. Difficulties Paying Bills

Some people find it difficult to manage risky events related to climate change, lack of cash, credit, land, networks, education, or time to adapt [120,121], leading to a lack of involvement in actions related to environmental protection. Significant linkage between household income and standard socio-demographic characteristics and environmental quality concerns has been demonstrated [122,123].

For this particular case, poor people or migrants tend to be the most vulnerable [124]. Regarding income, the evidence is inconclusive, the actions pro-environment is unrelated to income, according to Domene and Sauri [125] and positively correlated, according to Renwick and Archibald [126].

2.3.7. Urban Area

Taking account of the high level of population concentration in cities, these consume three-quarters of global energy consumption and greenhouse gas emissions [127]. That's why cities are fighting to mitigate climate change by implementing low-carbon development strategies [128]. Cities involve wealth and innovation, with resources and tools necessary to combat climate change [129].

Intercultural analyses of climate change and environmental protection behavior have attracted increasing attention recently [52–54], highlighting variations in environmental concerns and public protection at the regional level [55,58,69,102], including efforts to mitigate climate change [60]: coastal or riverine regions are strongly unprotected and thus vulnerable in facing climate change impact [130], which poses a huge pressure on urban infrastructure, the lives of urban citizens and the entire urban system.

Climate change mitigation is more likely in large cities than smaller cities [131], most often the focus being on energy policy, air quality and GHG emissions.

2.3.8. Political Ideology

Another important variable that influences environmental behavior is political ideology [48,64,88,99,110,112,132], and people with left-center political visions are more concerned and have significant environmental attitudes and behavior [96,112,113,133]. Political ideology has clear relationships with beliefs about climate change [134,135], and conservatives and right-wing citizens have low beliefs about climate change [136], being more skeptical on this issue [46]. The explanation is that conservatives have strong tendencies to justify the system, denying issues that threaten the functioning of the system, including climate change [137]. Important results have been provided by the study of Ortega-Egea et al. [5] who have proved that individual having a leftist/liberal political orientation were more likely to engage in extra mitigation behavior compared to right-wing/pro-environmental respondents.

2.3.9. Political Interest Index

Citizens' beliefs directly influenced their implication in the public policy and have not been the subject of many researches, as there are few studies in the literature on this connection [138]. Political distrust is more prevalent in countries with more corruption [86]. People who trust politics support environmental protection policies [72,139,140], and countries that register a high level of political trust are more inclined to spend for environmental protection [141]. In developed countries there is a tendency to diminish political trust [142], this being considered the reason for the low involvement of citizens in actions aimed at improving the common good [143]. On environmental issues, people do not trust governments, businesses, industry, and experts, and hold them accountable for controlling environmental risks [36].

According to the results of Fairbrother et al. [86], supported at the level of 23 European countries, higher political trust was related to a higher level of support in adopting fossil fuel taxes among individuals believing in the importance and awareness of climate change.

2.3.10. Modernisation Theory-Related Factors

The existent differences in pro-environmental beliefs are for sure related to the existent differences in wealth modernisation level [54,60,68], stating that citizens from richer nations grant a higher level of importance to all environmental targets compared with those from pooper countries. The environmental awareness is positively correlated with wealth [67], and negatively correlated with work [64,67,68] in poorer countries, generally the environment related issues are subjects of public concern. For individuals in rich countries, the general perception is that climate change is a fundamental issue and in consequence it is not ranked as a highly dangerous threat, leading to maladaptation to climate change [144].

Individuals of richer countries have this perception about climate change to be a fundamental issue and consequently this could lead to maladaptation to climate change [144–146].

Since *greenhouse gas* emissions significantly influence climate change, modernization theory and world economy theory [147] have been investigated, empirical results confirming both theories. The IPCC WGII [148] proved that the main factors of influence in the climate change adaption are the access to economic and natural resources as well as institutions and governance.

Well-being is very important in climate change mitigation, and human need and quality of life interferes with energy and resources usage within society [149–151]. Adua et al. [152] drew attention that the modern communities need to be redesigned if the target is to attenuate the environmental crisis [153] still registering economic growth. Between technological innovations indicators and CO_2 emissions was registered a negative relationship, with technological innovation being very important in diminishing human impacts on the environment.

Digitalization foster climate-friendly urban environments [154] being considered a stepping stone for the global economy [155].

ICTs (Information and Communication Technologies) brings a contribution to climate change mitigation through the access to relevant information, increasing concerns and making possible the enhancing of learning and knowledge [156]. Individuals who exhibit a higher level of information using diverse and multiple sources of information, [157] have also a different level of environmental issues' perception, including climate change.

In order to mitigate and adapt to climate change, digital technologies are very important, with information and communications technologies leading to cities climate action. Involving citizens and making possible their participation in city programs with appealing digital technology are very important for climate action [158]. Exploring similarities in environmental protection and climate change mitigation among Norway's voters, Kullberg and Aardal [159] revealed that those in favor of environmental protection and climate change mitigation possess rather a libertarian way of thinking compared with climate-only supporters who exhibit an authoritarian belief. The governmental policy related to climate change put in the first place mitigation instead of adaptation [160], taking into account on the one hand the main benefits of mitigation [161] as well as other factors such as savings, energy security or the importance of reducing emissions [162–166]. Governments significantly influence environmental issues, and can intervene politically through taxes related to the damage caused to the environment. Thus, polluters pay for each environmentally harmful action, discouraging pollution and encouraging companies to find and implement new technologies and production processes that affect the environment as little as possible [167,168].

Taking into account all these, the present research represents, to our knowledge, the first study exploring the specific mitigation behavior of climate change through the analysis of 13 types of personal actions and their determinants, as well as one of the few studies exploring the extra mitigation behavior determinants in a more comprehensive way.

Acknowledging the important role of personal behaviour in tackling climate change from different perspectives, supporting the shift to a circular economy, the following hypotheses have been formalised in order to highlight the main factors related to the level of citizens' specific mitigation and extra mitigation behaviour:

Hypothesis 1. (H1). *An increase in pro-environmental behaviour is associated with a higher level of engagement in adopting specific actions in fighting climate change as well as with a higher level of extra mitigation behaviour, supporting the switch to a circular economy.*

Hypothesis 2. (H2). *Individuals with the following socio-demographics characteristics (female, from 55–64 age group, married with children, with a higher level of education, self-employed, with a left- center political ideology and a higher level of political interest) are more inclined to engage in specific actions in fighting climate change as well as to exhibit a higher level of extra mitigation behaviour, supporting the transition to a circular economy.*

Hypothesis 3. (H3). *Citizens from small or medium towns, without facing any difficulties in paying their bills and living in Western Europe or Nordic nations are more likely to engage in specific actions in fighting climate change as well as to exhibit a higher level of extra mitigation behaviour, supporting the transition to a circular economy.*

Hypothesis 4. (H4). *Citizens from more developed and modernized countries are more likely to exhibit a higher level of extra mitigation behaviour, supporting the transition to a circular economy through the multiple actions undertaken.*

Therefore, a positive sign is expected, a higher level of SPI leads to a higher level of engagement in climate change mitigation.

The Happy Planet Index (HPI) reflects sustainable wellbeing for all individuals, quantifying the process of achieving a long, happy, sustainable lives, and puts together four dimensions: wellbeing, life expectancy, inequality of outcomes and ecological footprint, in order to quantify how citizens of different countries are using environmental resources to lead long, happy lives. Therefore, the expected sign is a positive one, a higher level of the index, a higher level of engagement in mitigating climate change.

Two proxies for digital usage and access have been taken into account—individuals' level of internet skills (individuals who have done at least one internet activity) in urban areas (%) and individuals' level of internet access in the last 12 months in urban areas (%); a positive impact being expected. Therefore, a higher level of individuals' internet skills and access is associated with a higher level of personal engagement in tackling climate change.

As governance indicators, we have included three main variables: SGI government effectiveness score, SGI policy performance score, SGI environmental policy protection score.

The first, SGI Government Effectiveness score is one of the SGI Governance indicators and measures the efficient implementation of governmental policies. Therefore, a positive impact is expected, a higher perceived level of government effectiveness being associated with a higher level of engagement in climate change mitigation.

The second one, SGI policy performance score, is another SGI Governance indicators measuring the government efficiency in cultivating sustainable conditions helping to assure well-being an empowerment. Also, a positive impact is expected, and a higher level of trust in government is associated with a higher level of personal mitigation of climate change.

The last one, the SGI environmental policy protection score, measures the level of resources protection and environment quality acquired through environmental policy, a positive impact being expected; a higher level of environmental policy protection being associated with a higher level of citizens' engagement in fighting climate change.

3.2. Multi-Level Econometric Modelling

In the process of highlighting the main driving forces of climate change among EU citizens, a multilevel logistic regression (the command in STATA 15 for the estimation of multilevel logistic regression models is xtmelogit) analysis has been used highlighting that the specific and extra mitigation behaviour is influenced by both individual characteristics and country specificities in combating climate change, supporting thus the shift to a circular economy.

The conceptual framework of the main determinants for citizens' engagement level in mitigating climate change is presented in Figure 3.

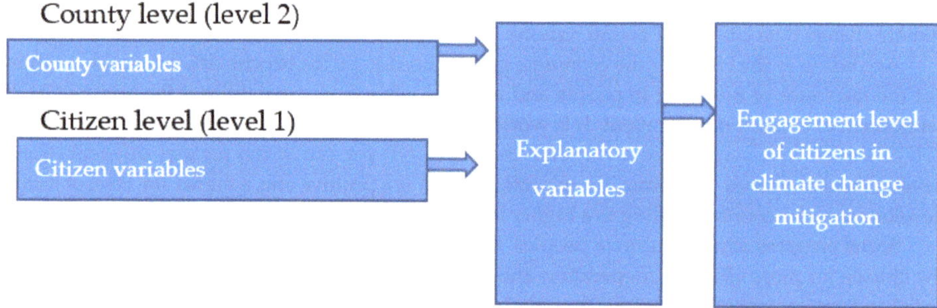

Figure 3. Conceptual framework of the main determinants for citizens' engagement level in mitigating climate change. Source: Author's contribution.

In order to analyse the between-country variation in the specific and extra mitigation behaviour of climate change, diverse categories of of two-level models were used. The two-level models implies a four-stage methodology.

In the first stage, the appropriateness of the multi-level approach was tested by the estimation of a baseline random intercept model without any explicative variables, the empty two level model with an intercept and country effects (the null model) has the following specification:

$$\log\left(\frac{\pi_{ij}}{1-\pi_{ij}}\right) = \beta_0 + u_{0j}, \quad (1)$$

The intercept β_0 is shared by all countries, while the random effect u_{0j} is specific to county j and it follows a normal distribution with variance σ^2_{uo}.

In the second stage, the model additionally assumes the influence of first-level characteristics:

$$\log\left(\frac{\pi_{ij}}{1-\pi_{ij}}\right) = \beta_0 + \beta_1 \cdot X_{ij} + u_j, \quad (2)$$

In the third step, the model specification includes simultaneously both individual-related factors as well as country-level variables being formulated as follows [85]:

$$\log\left(\frac{\pi_{ij}}{1-\pi_{ij}}\right) = \beta_0 + \beta_1 \cdot X_{ij} + \beta_2 \cdot X_j + u_j, \quad (3)$$

where: β_0 is the overall intercept, β_1 is the cluster specific effect, β_2 is the contextual effect, X_{ij} is the vector containing individual level explanatory variables and their interactions, X_j is the vector containing country level explanatory variables and u_j is the group (random) effect.

As in any regression model, we can include interaction effects which allow for the possibility that the effect of one explicative variable on the outcome depends on the value of another explanatory variable. An interaction between a level 1 variable and a level 2 variable is called a 'cross-level interaction'. Furthermore, it is worth testing if the effect of contextual country factors on climate change extra mitigation behavior depend on the type of the town (large town vs. small and medium size town). Therefore, in the fourth step, different random intercept models with cross-level interactions have been estimated.

$$\log\left(\frac{\pi_{ij}}{1-\pi_{ij}}\right) = \beta_0 + \beta_1 \cdot X_{ij} + \beta_2 \cdot X_j + \beta_3 \cdot X_{ij} \cdot X_{2j} + u_j, \quad (4)$$

The general scheme of two level models for citizens' engagement level in combating climate change is displayed in Figure 4. Different specifications of two-level logistic models were estimated in over to overcome the issue of multi-collinearity, as country-level variables could be correlated.

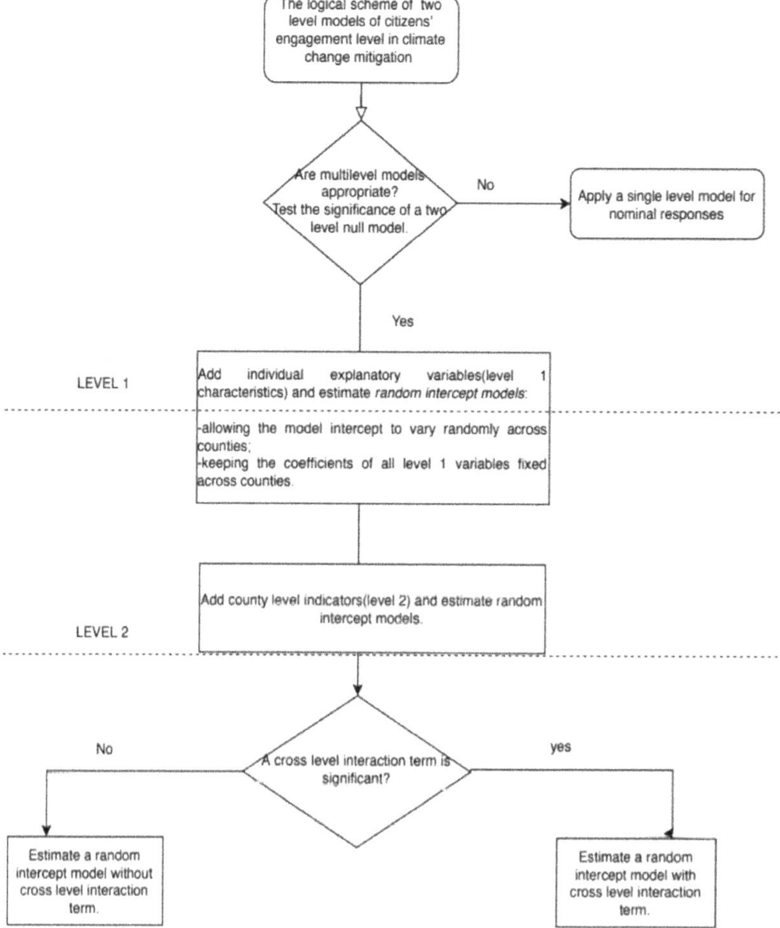

Figure 4. The logical scheme of two-level models for climate change mitigation behaviour. Source: Author's contribution.

4. Empirical Results and Discussion

4.1. Descriptive Profile of Respondents

From 18,529 individuals from the urban environment, almost 42.7% of them come from a large town while 57.3% of respondents live in small and medium-sized towns. Also, 38.66% of them were citizens of East-Central Europe and only 13.38% come from Nordic nations.

The sample is well-proportioned in terms of gender, age, education; yet, there was greater involvement of female (54.3%), middle-aged (mean age = 51.09 years), and moderately educated individuals (41.84% of them finishing school at the age of, at most, 19 years old) and 34.81% have multiple marriages without children, a centre political orientation (42.08%). Almost 47.33% of them were not working, 45.84% of them were employed and only 6.84% were self-employed and 66.91% of respondents declared that never had difficulties in paying bills and only 7.8% of them declared that most of the time faced this issue.

Almost half of respondents have a medium level of political interest and only 16% of them were not at all interested in such a topic.

Asked about the importance of climate change, an overwhelming proportion of 78.71% of them perceived it as a very serious problem and only 6% of individuals considered it not a big problem. Also about 88.7% of citizens agreed with the fact that action on climate change leads to innovation making companies more competitive.

In the descending order of perceived responsibility, most citizens considered national governments responsible for the mitigation of climate change (55.28%), followed by business and industry (52.12%), European Union (49.22%), citizens (35.41%), regional and local authorities (33.09%), and environmental groups (29.04%). All of these actors have been mentioned only by almost 11% of the respondents. Asked if they undertook specific actions in fighting climate change in the last six months, almost 63% of respondents consented to this statement.

4.2. A Descriptive Analysis of Climate Change Mitigation among European Union (EU) Countries

The climate change was perceived as a very serious problem by most of the respondents in each country, with an overwhelming proportion of citizens in Malta (93%), Greece (92%) and Spain (89.5%). Only a very small proportion of citizens from every country considered that the climate change is not a serious problem, with two-digit percentages for Finland (11.6%) and Austria (16%).

Asked about how the main responsibility of tackling climate change is divided among the main actors in the field, the majority recognised the important role of national governments, EU institutions as well as business and industry. More than half of all respondents from urban areas attributed the main responsibility to national governments (55.3%), business and industry (52.1%) as well as EU instructions (49.2%). Just over 30% of respondents recognised the role of themselves as important drivers in tackling climate change, as well as the contribution of local authorities. Almost 29% of respondents mentioned environmental groups as being responsible for tackling climate change. What it is interesting to mention is that only 11% of all respondents attributed the responsibility of combatting climate change to all actors involved.

From all EU member states, 18 countries recognised the importance of national governments' responsibility in combating climate change, the leaders being Sweden (80.2%), the Netherlands (73.6%), Malta (73.4%) and Greece (67.8%), while in the vision of the respondents from Sweden (57%), Netherlands (57%) and Finland (51%), citizens were seen as a responsible body in tackling climate change.

More than 4 out of 10 respondents in Sweden and Portugal (47%, respectively, 43%), Romania (44%) and Malta (42%) assigned the responsibility of combating climate change to regional and local authorities.

An interesting remark is related to the fact that 18 EU countries attributed the responsibility of combatting climate change to all actors involved, with the highest proportions in Portugal (21.6%), Slovenia (19.7%) and Luxembourg (19.6%).

Citizens from the urban areas of more than 22 countries declared that they have personally taken action to fight climate change in the past six months and this is available mostly in the case of Slovenia (86%), Malta (85%), Sweden (81%), Spain and Finland (80%) and Luxembourg (78%) compared to only 31% in Romania, 34% in Bulgaria and 40% in Poland (Table A2, Appendix A).

Asked about specific actions in fighting climate change (Figure 5), 73.4% of respondents declared that they tried *to reduce their waste and regularly separate it for recycling*, while 62% of them resorted to *reducing consumption of disposable items*, these being the most common measures taken to combat climate change. The Energy EFF HH appliances was ranked third in the list of measures taken to mitigate climate change, these three actions being considered as the most common measures taken by more than half of the respondents.

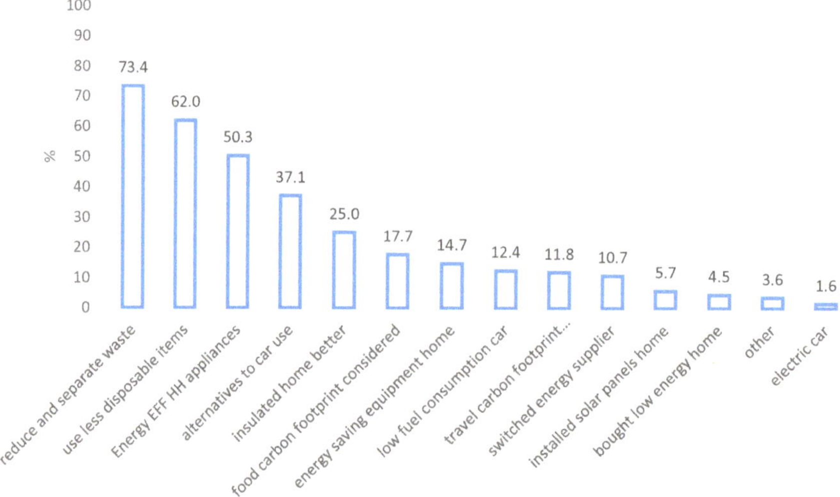

Figure 5. Types of individual action taken to fight climate change. Source: Author's contribution.

In order to combat climate change, about 37% of the respondents declared using *environmentally-friendly alternatives to their private car*, while 25% of them *insulated their home better.*

A small proportion of citizens, almost one in five (18%) take into account *the carbon footprint of their food purchases* and tried to accommodate their shopping accordingly, while an even lower percentage of interviewed citizens (14.7%) declared that they *installed energy-saving equipment home* and only 12.4% of respondents admitted that they have tried to reduce *the fuel consumption buying a new car.*

The measure of considering the carbon footprint of transport, adapting their holiday accordingly, has been adopted by only 11.8% of the respondents, while an even lower proportion of individuals claimed that they have *switched energy supplier in order to increase the energy share from renewable sources* (10.7%). Installing solar panels (6%), buying a low-energy home (5%) or buying an electric car (1.6%) are the least undertaken actions among EU citizens.

In all countries, except for two, the majority of respondents declare they try undertook the action of reducing and separating waste. The country recording the highest proportion of reduced and separate waste is Sweden (91.8%), followed by Netherlands (87.8%), France (87.7%) and Luxembourg (87.5%), while the majority of respondents declaring to *reduce consumption of disposable items* were from Netherlands (84.8%), Sweden (84%) and Luxembourg (78.2%).

The energy EFF HH appliances measure taken to reduce energy consumption has been the respondents' choice most in the Netherlands (77.6%), Germany (65.6%) and Latvia (64.9%), while regarding the alternatives to car use, the first three countries adopting this measure have been Sweden (67.5%), the Netherlands (67.1%) and Germany (55.9%).

Most respondents considering to insulate their home better in order to reduce their energy consumption were from the Netherlands (39.6%), Bulgaria (36.8%) and Denmark (36%) while Finland and Sweden (41.3%) were the countries registering the biggest proportion of urban citizens that considered the carbon footprint of their food purchases.

The share of citizens who bought a new car, thus reducing fuel consumption, was higher in Denmark (27%), Sweden (20.1%) and Belgium (18.4%) while the highest adoption rate of considering travel carbon footprint when planning their holiday have been encountered among citizens from Sweden (41%) and Finland (26%).

An important measure for climate change mitigation is switching to an energy supplier, contributing thus to an increase in the energy share from renewable sources. In this regard, Sweden registered an adoption rate of 26.8%, the Netherlands (24.5%) and Belgium (22.9%).

Urban respondents having installed solar panels in their home are most from Cyprus (24.4%) and the Netherlands (21.5%), while 12.7% of citizens in Netherlands and 12.5% of citizens in Luxembourg have bought a low-energy home.

Overall, the countries registering the biggest proportion of measures mitigating climate change were Denmark (27%), Netherlands (20.7%) and Sweden (20.1%). The countries registering the lowest proportion of measures mitigating climate change were Greece (4.6%), Spain (4.4%), Portugal (4.4%) and Poland (4.7%).

Analysing the distribution of the number of individual actions taken to combat climate change it can be highlighted that less than 10% of EU urban citizens took no measure to mitigate climate change, 25% took one or two actions, and 50% took at most three actions.

Only 25% of the respondents declared that they have taken more than five actions and only 1% of individuals declared to taken 9 or more actions. In urban areas, EU citizens took, on average 3 measures, the maximum of the measures taken being 13 (Table 1).

Table 1. Summary statistics on action to mitigate climate change.

Percentiles		**Smallest**		
1%	0	0		
5%	0	0		
10%	1	0	Obs.	18.529
25%	2	0	Sum. Of Wgt.	18.529
50%	3		Mean	3268
		Largest	Std. Dev.	2165
75%	5	13		
90%	6	13	Variance	4688
95%	7	13	Skewness	6519
99%	9	13	Kurtosis	3229

Regarding the cumulative measures to mitigate climate change, it can be seen that the majority from EU urban areas (91.85) took at least one measure, 77.4% of EU urban citizens took at least two actions and only 15.4% of EU urban citizens took six or more actions. The Netherlands is the country whose citizens adopted more than six actions in fighting climate change (52.9% in case of 6 measures) while at the opposite end there is Poland, with only 1.1% of citizens adopting six actions in combating climate change.

4.3. Results of the Estimated Models

In accordance with our first objective, the research explored the main socio-demographic factors related to specific mitigation behaviour quantified by individual actions untaken to fight climate change, with the estimation process including 13 models, one for each type of climate change individual action.

The empirical results of the null baseline random intercept model, revealed that multilevel approach was appropriated and the log-odds of citizens share declaring having undertaken specific actions to fight climate change in an "average" country was estimated to be $\beta_0 = 0.60$.

The between-country variance of the log-odds of citizens' share taking actions to combat climate change was estimated as 0.455 with a standard error of 0.125. The significance of the between-region variance was tested using a Wald test, the empirical results (Chi-square test = 1456.74 with a p-value = 0.0099) revealing that there is a significant variation between EU countries in the proportion of those citizens environmentally active taking specific actions in fighting climate change.

Based on the value of between-country variance (0.455), the variance partition coefficient (VPC) was computed to be 12.15% using the formula: 0.455/(0.455 + 3.29) = 0.1215; thus 12.15% of the residual variation in the propensity of taking actions in mitigating climate change is attributable to unobserved country characteristics), indicating that almost 13% of the variance in climate change specific mitigation actions can be attributed to differences between countries.

Figure A1, Appendix A, reports the residual level-2 country effects derived from the null model (with no explanatory characteristics) proving the differences between countries. A country whose confidence interval does not overlap the line at zero (representing the mean log-odds of taking specific actions to climate change mitigation across all states) differs significantly from the EU-28 average at a 95% confidence level [84].

Romania, Bulgaria and Poland are countries with the lowest probability of taking personal actions in mitigating climate change in the last six months (largest negative values of u_j) for which the confidence intervals do not overlap with 0, pointing out that a significantly lower probability of participation compared with the EU average has been registered by these countries. At the upper end, Sweden, Slovenia and Malta are the countries with intervals that do not overlap with 0 with the highest response probability (largest positive values of u_j), indicating a significantly higher probability of participating in climate change actions compared with the EU-28 average.

Proving that the multilevel approach is the appropriate one, in the second stage different specifications of the multilevel models with only individual level characteristics were stipulated, with 13 sequential models being estimated, one for each specific action taken by citizens in the mitigation process of climate change, providing new and more detailed perspectives regarding the main determinants of climate change specific mitigation behaviour. The empirical results of random intercept models including individual-level variables are presented in Table A3, Appendix A.

For the first specific climate change action adopted at the same time by most of the citizens, *reduce and separate waste,* the empirical results pointed out the statistical significance of climate change perceived importance, the perceived responsibility of national governments, European Union, business and industry, citizens and environmental groups, as well as the importance of climate change actions leading to innovation in explaining the specific mitigation behaviour.

Also, most of the socio-demographic variables exhibited statistical significance: gender, age, education, occupation, facing difficulties in paying bills, marital status (children), type of urban area, types of EU region, political interest as well as political ideology.

Analysing the individual characteristics, it can be highlighted that women, from the age groups 45–54 years and 55–64 years, with a high level of education, with multiple marriages with children, usually employees, without difficulties in paying their bills, living in small towns from Western Europe, Southern Europe or Nordic Nations, having a leftist/liberal or centre political orientation and a high level of political interest, were inclined to adopt this specific action in fighting climate change.

The empirical results of the second specific action adopted by the most respondents—*use fewer disposable items*—revealed that people perceiving the climate change as a serious problem, considering that the responsible bodies in fighting this phenomenon are national governments, the European Union, business sector, local authorities or citizens themselves or acknowledging that the role of climate change actions are more inclined to undertake this action. Analysing the profile of individuals adopting this action, it can be mentioned that women more than 35 years old, with a high level of education, with multiple marriages with children, usually without any problem in paying bills, living in Western Europe and Nordic nations having a leftist/liberal or centre political orientation and a high level of political interest were more inclined to fight climate change using less disposable items.

The empirical results of the third action taken to combat climate change—*energy EFF HH appliances*—highlighted that people perceiving climate change as a very serious problem, those considering that the main responsibility in tackling this negative phenomenon rests with national governments, European Union, business sector or themselves, as well as those acknowledging the importance of climate change actions on the path to innovation, were more willing to adopt this action.

The fourth action adopted in combating climate change—*alternatives to car use*—was more likely to be considered by people perceiving the climate change as a very serious problem, those considering that the main responsibility in tackling this negative phenomenon rests with national governments, European Union, or themselves. Regarding individual characteristics, more educated women, 55 years old and over, living usually in large towns, having multiple marriages without children, and without having difficulties in paying bills or living in Western Europe as well as in Nordic nations, having an increased level of political interest and a leftist/liberal political orientation, are more determined to combat climate change by adopting this specific measure.

The fifth action mentioned by the citizens in the climate change mitigation— *better insulated home*—was more likely to be considered by people perceiving the climate change as a very serious problem, those considering that the main responsibility in tackling this negative phenomenon returns mainly to citizens. Exploring the citizens' profile, it can be highlighted that men, with a higher level of education, from all age categories, usually self-employed, living in small towns, having multiple marriages with or without children, without any difficulties in paying bills and having an increased level of political interest and a right/conservative political orientation.

Concerning the sixth action adopted—*food carbon footprint*—the empirical results revealed that people perceiving the climate change as a very serious problem, those considering that the main responsibility rests mainly with national governments, regional or local authorities as well as to citizens, or acknowledging the importance of climate change actions on the path to innovation were more likely to take such an action. Highlighting the individual profile, it can be mentioned that women, from the age category 65 years old, more educated, having multiple marriages with our without children, not facing any difficulties in paying bills, living in Western Europe or Nordic nations, having an increased level of political interest or a leftist/liberal political orientation, were more inclined to adopt this action as climate change mitigation measure.

The empirical results of the seventh action—*energy saving equipment at home*—revealed that citizens considering that the main responsibility returns regional or local authorities, environmental groups or even themselves and living in small towns from Western Europe or Nordic nations, from all age categories, with a higher level of education, usually self-employed, having multiple marriages with or without children, with a very strong political interest and a liberal political orientation, were more inclined to adopt this action.

The eighth action—*low fuel consumption car*—was taken as climate change mitigation more likely by those citizens perceiving climate change as a very serious problem, attributing the main responsibility of climate change mitigation to citizens, those living in small towns, from countries in Western Europe or Nordic nations, without any difficulties in paying bills, being men, either less than 44 years old or over 65 years, with a higher level of education, usually self-employed, having multiple marriages with our without children, having a medium and strong level of political interest and a liberal political orientation.

The ninth action—*travel carbon footprint*—was more likely by those citizens perceiving climate change as a very serious problem, attributing the main responsibility of climate change mitigation to citizens, living in large towns, from Western Europe or Nordic nations, not facing any difficulties in paying bills, being women, educated people, usually self-employed, having a medium and strong level of political interest and with a leftist/liberal political orientation.

The empirical results of the tenth action—*switched energy supplier*—revealed that those citizens perceiving climate change as a very serious problem, considering that the main responsibility rests with citizens, acknowledging the importance of climate change actions on the path to innovation, living in Western Europe or Nordic nations, with a high level of education, usually self-employed, having multiple marriages with our without children, an increased level of political interest, were more inclined to adopt this action in the mitigation process of climate change.

The eleventh action—*installed solar panels at home*—was a climate change mitigation more likely to be taken by those citizens considering that the main responsibility returns to regional authorities, business sector, living in small towns, from Western and Southern Europe, educated people, from the age category 55–64 years old, usually self-employed, having multiple marriages with our without children and a centre political orientation.

The twelfth action—*bought low-energy home*—was a climate change mitigation more likely to be taken by those citizens considering that the main responsibility returns to business sector, living in small towns, from Western Europe, being men, educated people, usually self-employed, not facing any difficulties in paying bills, having multiple marriages with our without children and a very strong political interest.

The last action—*electric car*—was a climate change mitigation more likely to be taken by those citizens considering that the main responsibility returns to regional and local authorities, living in Western Europe or Nordic nations, educated people, having at most 44 years old or over 65 years old, usually self-employed, having multiple marriages with children and a right/conservative political ideology.

The mitigation actions exhibiting the largest variance at country level among individual characteristics were the following: installed solar panels home (12.27%), reduce and separate waste (10.59%), switched energy supplier (9.37%) and food carbon footprint (9.12%).

Going further in our research analysis, in accordance with our second objective and after exploring the main determinants of citizens' decisions to undertake specific individual actions to fight climate change, the research investigated factors that relate to extra behavioural engagement, trying to find what determines some people to be involved in an extra commitment (i.e., do "more") to mitigate climate change as compared to others, testing the validity of modernization thesis. In this context, the modernisation thesis has been tested, stipulating that extra mitigation behaviour is more likely to occur in more modernized developed economies. Therefore, in order to prove this, 12 alternative models have been estimated, for different proxies of modernization thesis including also those models taking into account cross-level interaction terms.

The empirical results of the null baseline random intercept model revealed that the multilevel approach was appropriated and the log-odds of extra personal engagement in mitigation behaviour in an "average" country was estimated to be $\beta_0 = -1.226$.

The between-country variance of the log-odds of citizens' extra mitigation behaviour was estimated as 0.842 with a standard error of 0.228. The significance of the between-region variance tested with the Wald test (Chi-square test = 2495.02 with a p-value = 0.00) revealed that there is a significant variation between EU countries in the extra personal engagement in mitigation behaviour.

The variance partition coefficient (VPC), determined using the value of between-country variance is 20.37%, revealing that 20.37% of the residual variation in the propensity of extra mitigation behaviour is attributable to unobserved country characteristics.

Figure A2, Appendix A reports the residual level-2 country effects derived from the null model (with no explanatory characteristics) proving the differences between countries. A country whose confidence interval does not overlap the line at zero (representing the mean log-odds of taking multiple actions to climate change mitigation across all states) differs significantly from the EU-28 average at the 5% significance level [83].

Poland, Croatia, Lithuania and Italy are countries with the lowest probability of extra mitigation behaviour (largest negative values of u_j) for which the confidence intervals do not overlap with 0, indicating that they have significantly lower probability of extra mitigating climate change than the EU average. At the upper end, the Netherlands, Sweden, Denmark and Luxembourg are the countries with intervals that do not overlap with 0 with the highest response probability (largest positive values of u_j), indicating a significantly higher probability of extra mitigation of climate change compared with the EU-28 average.

Stipulating that the multilevel approach is the appropriate one, the research built different specifications of multilevel modelling including both individual level characteristics as well as country-level factors related to modernisation thesis in order to extra mitigate climate change.

Therefore, 12 alternative models have been estimated capturing the main determinants of extra mitigation behaviour among EU citizens, the empirical results being presented in Table A4, Appendix A.

The empirical results of all alternative models revealed that climate change extra mitigation personal behaviour was more likely to be adopted by citizens perceiving the climate change as is a serious problem, attributing the responsibility of fighting against this phenomenon to national governments, regional or local authorities, business sector or themselves contributing through multiple actions undertaken to the transition to a circular economy.

Also, the importance of climate change actions leading to innovation and making companies more competitive was positively associated in all models with an extra personal engagement in mitigation behaviour.

All tested socio-demographic variables manifested a statistically significant impact on the decision of adopting an extra mitigation personal behaviour, the estimates of individual-level demographics being almost identical in all alternative models.

Female gender, from the age category 55–64 years old exhibited a higher level of extra mitigation habits, the empirical results pointing out the existence of an inverted U-shape between age and citizens' extra behavioural engagement in fighting climate change. Also, a positive association has been found between education and the extra mitigation behaviour, with people with a higher level of education being more inclined to adopt multiple actions manifesting a higher level of commitment.

Also, the empirical results revealed that people of a leftist/liberal political orientation and a higher level of political interest displayed an extra behavioural engagement in climate change mitigation, while self-employed people, without any difficulties in paying bills, married with or without children, living in small towns, in Western Europe or Nordic nations were more willing to engage in extra mitigation behaviour.

The modernisation thesis have been only partially supported by the empirical results of models II, IV, V, VI, VIII, IX; a higher level of extra mitigation behaviour being encountered among citizens from countries with a higher level of human development and social progress, a higher proportion of individuals with internet skills and access, and a higher level of policy performance and environmental policy protection.

Therefore, the modernisation thesis was strongly supported by the results of the second model; citizens from countries with a higher level of human development exhibiting also a higher level of extra behavioural engagement to mitigate climate change. A similar result was confirmed by the statistical significance of social progress index; citizens from countries with a higher level of social progress exhibited a higher level of extra engagement to mitigate climate change. Also, the digitalisation hypothesis made its contribution to the validity of the modernisation thesis, with citizens from countries with a higher level of individual internet skills and access being more determined to engage in extra mitigation behaviour.

The modernisation thesis was also supported by the statistical significance of SGI governance indicators; citizens from countries with a higher policy performance as well as those from countries with a higher environmental policy protection were more willing to adopt multiple actions in fighting climate change.

Finally, it can be pointed out that the modernisation thesis was only partially validated, due to the lack of statistical significance of GDP per capita at PPS, a proxy for a country's wealth level; the HPI index a proxy for the level of country sustainable well-being; and the SGI government effectiveness a proxy for the efficient implementation of governmental policies.

Furthermore, it was worth testing if the effect of modernization thesis on the extra behavioral engagement to mitigate climate change, depends on the segmentation large town vs. small or medium town. Therefore, different random intercept models with cross-level interactions have been estimated.

Until now, the models assumed that the contextual effect of the modernization thesis is the same for all citizens regardless of whether they live in big or small cities. In this stage, we have modified this assumption allowing the effect of modernization thesis on the extra mitigation behavior to depend on the country level variables associated with this thesis and also on the urban area type, including in the model the interaction between country level variables and the type of the town in which the citizen is living as cross-level interactions terms. Another nine alternative models have been estimated, but given the limited space within the paper, we have decided to present only those exhibiting statistically significant impact on the extra mitigation behavior.

The empirical results of the models X, XI and XII displayed in Table A4, Appendix A revealed that the marginal effect of modernization thesis supported by the statistical significance of HDI, SPI, SGI environmental policy protection score on the level of extra behavioural engagement to mitigate climate change positively and significantly depends on the segmentation large vs. small or medium town.

Therefore, in a country with a higher level of human development, social progress or a higher level of environmental policy protection, citizens living in large towns exhibited a higher level of extra engagement in mitigating climate change compared with those living in small or medium towns.

4.4. Summary of Findings in Relation to Specific Mitigation and Extra Mitigation Behavior

In Figure 6, we summarize our overall findings in relation to the main hypotheses tested, providing information on the positive relationship between specific factors and each specific action of combating climate change as well as an extra mitigation behaviour in fighting climate change. It also provides information on the modernisation theory supported by each statistically significant finding, along with clarifying remarks.

Analysing the empirical results of specific mitigation behaviour models, the pro-environmental variables were statistically significant in eight of the specific actions models, stipulating the relevance of pro-environmental attitudes in combating climate change. Therefore, individuals acknowledging the importance of climate change, being aware of their responsibility in fighting this phenomenon together with other institutional bodies, and recognising the importance of climate change actions were found to be more inclined to adopt specific actions in mitigating climate change. Thus, hypothesis H1 has been partially validated for the specific actions: reduce and separate waste, use disposable items, energy appliances, alternatives to car use, insulated home better, food carbon footprint, low fuel consumption car, switch energy supplier. Also, hypothesis H1 has been fully validated in the case of extra mitigation behaviour, by the positive and statistically significant coefficients of pro-environmental proxies.

Hypothesis H2 regarding the profile of individuals engaged in specific mitigation behaviour is validated in most of the cases, with the exception of three specific actions. Therefore, females, mostly 55–64 years old, married with children, with a high level of education, self-employed, with a left-or center political ideology and a high level of political interest) were more inclined to adopt the following specific actions as measures to combat climate change: reduce and separate waste, use disposable items, energy appliances, alternatives to car use, food carbon footprint, switch energy supplier, energy-saving equipment home, travel carbon footprint, electric care. For the other specific actions concerning insulated home better, low fuel consumption car and bought a low-energy home, the profile of individuals inclined to adopt these actions is: male, young people (15–24 years old), with high level of education, self-employed, married with or without children, having a right political ideology and a high level of political interest.

Hypothesis H2 has been fully validated in the case of extra mitigation behaviour, by the positive and statistically significant coefficients of socio-demographic variables.

Hypothesis H3 has been partially validated by the positive and statistically significant coefficients of the models investigating the following seven specific actions: reduce and separate waste, use disposable items, alternatives to car use, food carbon footprint, low fuel consumption car, travel carbon footprint and bought low-energy home. Therefore, individuals not facing any difficulty in paying bills, living in small or medium towns from Western Europe or Nordic nations are more likely to engage in specific actions in fighting climate change.

Hypothesis H3 has been fully validated in the case of extra mitigation behaviour by the statistically significant coefficients of economic variables.

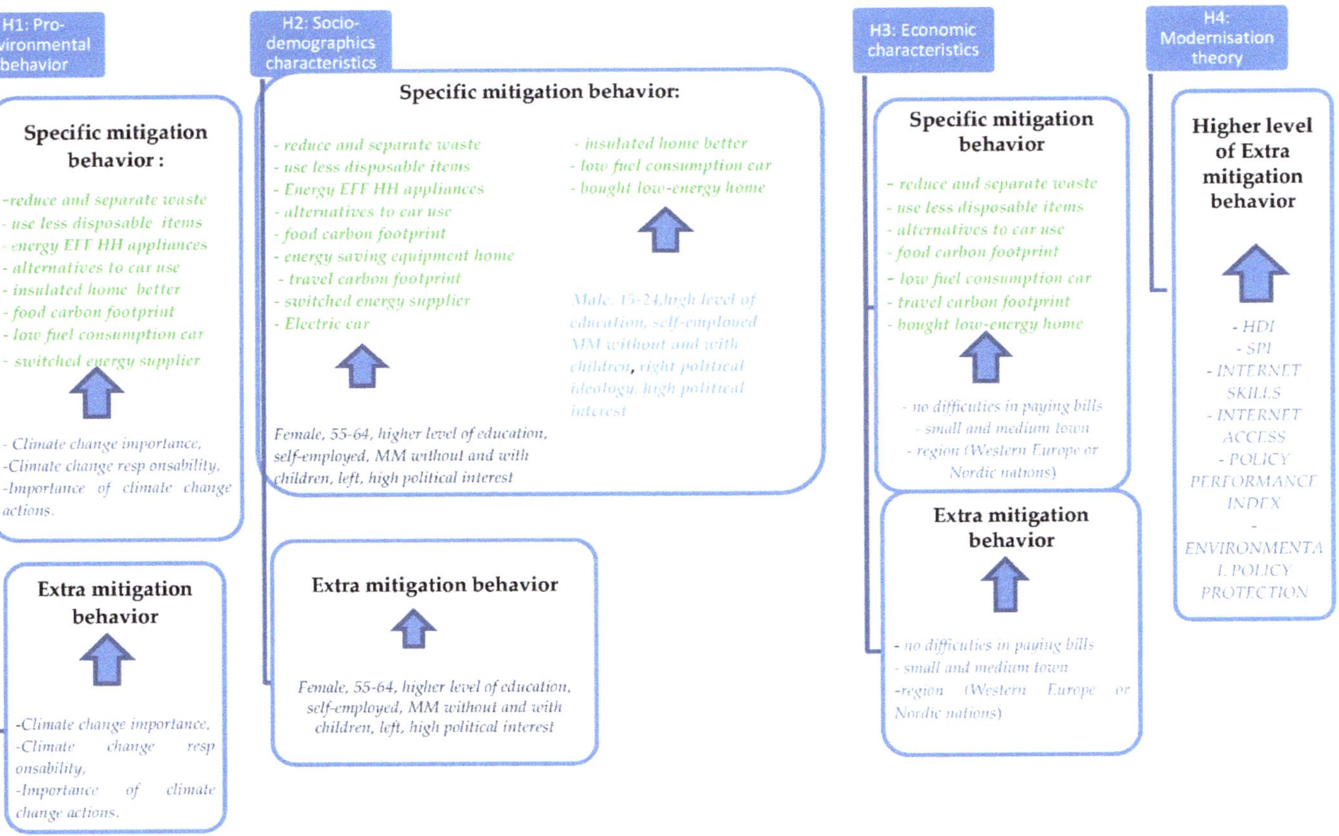

Figure 6. Summary of findings in relation to specific mitigation and extra mitigation behavior.

In the context of urban environments, it was important to analyse the impact of the segmentation small-medium urban area vs. large urban area on the decision of specific mitigation behaviour in fighting climate change, the empirical results clearly pointing out the statistically significance of urban area coefficients, revealing also that actions such as energy EFF HH appliances, insulated home better, energy-saving equipment home, low fuel consumption car, installed solar panels home, bought low energy home, were more likely to be adopted by citizens from small towns.

Hypothesis H4 concerning the validation of the modernisation thesis in relation to an extra mitigation behaviour has been only partially supported by the positive and statistically significant coefficients of human development index, social progress index, individuals' level of internet skills and access, SGI policy performance score and SGI environmental policy protection score. Therefore, citizens from more developed and modernized countries are more likely to exhibit a higher level of extra mitigation behaviour.

A related finding acknowledged that the effect of modernization thesis on extra mitigation behaviour depends on the segmentation large town vs. small or medium town, this fact being supported by the positive and highly significant coefficients of cross-level interaction terms between large town and country level variables related to modernisation thesis, revealing that citizens living in large towns in countries with a higher level of human development, a higher level of social progress or a higher level of environmental policy protection, were more likely to exhibit a higher level of extra mitigation behaviour.

4.5. Discussion

In order to mitigate climate change and also to pave the way to circular economy, the role of citizens is fundamental and a shift in personal behaviour is highly required. In combating climate change, the actions need to cover multiple measures and to target the implications of the main actors-local governments, business sector, civic society, environmental organisations and European institutions. The present research acknowledges the important role of citizens in tackling climate change, aiming to assess the role of socio-demographic, economic and environmental factors in fighting climate change as well as the validity of the modernisation thesis regarding the extra mitigation behaviour, exploring the between-country variation in both specific and extra mitigation behaviour.

A central question in this study was: what determines some people to engage in adopting specific and extra mitigation behaviour compared to others? The findings revealed that there is a mix of factors that influence both specific and extra mitigation behavior.

Overall, the findings reinforce earlier evidence that pro-environmental behaviour acknowledging the importance of climate change, climate change actions and the main responsibility in tackling this phenomenon are more associated both with a specific mitigation as well as an extra mitigation behaviour [58,59,62,67,69].

The perceived importance of climate change was significantly associated with specific and extra mitigation behaviour. In line with the previous studies [63–67], the findings emphasize the role of perceived importance of climate change on eight types of specific mitigation actions as well as on extra mitigation behavior.

Another driver of specific and extra mitigation behavior was related to the perceived responsibility in tackling this phenomenon, with the empirical results subscribing to studies in literature [87]; if citizens are more aware about their own responsibility and actions in mitigating climate change, this awareness will lead them in the direction of acting and adopting specific or multiple actions in fighting this phenomenon.

In contract with previous studies [91,92], the perceived importance of climate change actions seems not to determine as expected individuals engaging in specific actions, with the empirical results revealing a significant impact only for few specific measures. This is not the case for extra mitigation behavior, in which the increased perceived importance of climate change actions leading to innovation is associated with a higher level of extra mitigation.

The research strongly supports the idea that-despite empirical evidence from previous environmental studies [80,98] a variety of individual and country-level socio-demographics are significant correlates of specific and extra mitigation behavior in response to climate change.

In line with the work of [156,158,159,161,168], partial validation of modernization thesis have been achieved-a higher level of engagement to mitigate climate change being predominant in countries with a higher level of human development and social progress, a higher proportion of an individual's internet skills and access, a higher level of policy performance and environmental policy protection.

Another important result of our analysis is related to the fact that the effect of modernisation thesis on the extra mitigation behaviour of climate change depends on the type of urban area. Therefore, citizens living in large towns in countries with a higher level of human development, a higher level of social progress or a higher level of environmental policy protection were more likely to exhibit a higher level of extra mitigation behaviour [162].

4.6. Limitations and Recommendations for Further Study

The study presents several limitations that should be taken into account for further research. Firstly, the use of Eurobarometer dataset, despite the advantages, restricted the aim of the research questions by not allowing the investigation of other relevant associations of mitigation behaviour. Thus, further research could incorporate additional factors among which the individuals perceptions of self-relevance or the split of environmental activism and citizenship into public/political. Secondly, the research used self-reported data (instead of actual measured mitigation behaviour). Thus, the discrepancy between perceived and objective knowledge (and their relationships) deserves future attention in studies on climate change-motivated behavior. Thirdly, the analysis took into account only environmentally active citizens from urban areas, due to the main purpose of our research regarding specific and extra mitigation behavior. However, the exclusion of the environmentally inactive respondents should be recognised as an important limitation and cautions should be attributed in generalizing the results to non-environmentally active individuals.

Future research could be enlarged by including a broader segment of environmental respondents (the least receptive to the problem of the environment). Fourthly the study focused only on direct associations between factors and extra mitigation behavior. Future research could explore also a potential mediation analysis between different factors in explaining better the mitigation behaviour. Also, in further work, the lagged influence of climate change importance in revealing the determinants of mitigation behaviour could be incorporated and tested. Last but not least, the validation of the modernisation thesis opens the door for exploring also other theses with impacts on climate change mitigation behaviour. In the near future, country differences could be analysed also from the perspective of individualism thesis, long-term orientation or regional discrepancies.

5. Concluding Remarks

Taking into account all the findings of our research, the following can be concluded:

1. A shift in personal behaviour is required—from limited or sporadic actions toward exhibiting an extra mitigation behaviour in fighting climate change.
2. Two types of behaviour can be encountered among EU citizens—an individual mitigation behaviour based on singular actions undertaken to combat climate change and an extra mitigation behaviour based on multiple actions taken in fighting climate change.
3. Pro-environmental factors, socio-demographics, economic and country-level factors proved their role in explaining both specific and extra mitigation behaviour through personal engagement.
4. Modernisation thesis substantiated its validity revealing that both specific and extra mitigation behaviour is higher in more modernised countries.

5. By acknowledging the role of citizens in addressing climate change, and encouraging them to undertake different specific actions and doing more in fighting climate change by adopting multiple actions, we pave the way for a circular economy.
6. Through their contribution to waste reduction, eco-shopping, eco-friendly transportation or domestic energy reduction, citizens ensure the transition to a circular urban economy.

Author Contributions: All authors have been equally contributed to the paper. All authors have read and agreed to the published version of the manuscript.

Funding: This research received no external funding.

Conflicts of Interest: The authors declare no conflict of interest.

Appendix A

Table A1. Variables explained.

Country-Level Explanatory Variables	Data Source	Expected Sign
	The modernisation theory	
Gross Domestic Product (GDP) per capita in PPS is computed to by the GDP in PPS (current prices, million purchasing power standards)/population.	National Ac. Database, Eurostat; Employment and Unemployment Database, Eurostat.	Positive
Human Development Index (HDI) 2019	UNDP, Human Development Reports, http://hdr.undp.org/en/content/human-development-index-hdi	Positive
Happy Planet Index (HPI) 2019	http://happyplanetindex.org/	Positive
Social Progress Index (SPI) 2019	https://www.socialprogress.org/	Positive
Individuals' level of internet skills in urban area (%) The share of individuals from urban area who have used a search engine to find information (cities, towns and suburbs).	https://appsso.eurostat.ec.europa.eu/nui/submitViewTableAction.do	Positive
Individuals' level of internet access in urban area (%) The share of individuals from urban area (cities, towns and suburbs) who have used the internet in the last 12 months.	https://appsso.eurostat.ec.europa.eu/nui/submitViewTableAction.do	Positive
Government Effectiveness Score 2019 from SGI Indicators	Sustainable Governance Indicators, SGI Network, https://www.sgi-network.org/2019/Governance/Executive_Capacity/Implementation	Positive
Policy performance score 2019 from SGI Indicators	Sustainable Governance Indicators, SGI Network, https://www.sgi-network.org/2019/Policy_Performance	Positive
SGI environmental policy protection score 2019 from SGI Indicators	Sustainable Governance Indicators, SGI Network, https://www.sgi-network.org/2019/Policy_Performance/Environmental_Policies/Environment/Environmental_Policy	Positive

Table A2. Individuals' climate change actions.

Country ID	Reduce and Separate Waste	Use Less Disposable Items	Energy EFF HH Appliances	Alternatives to Car Use	Insulated Home Better	Food Carbon Footprint	Energy Saving Equipment Home	Low Fuel Consumption Car	Travel Carbon Footprint	Switched Energy Supplier	Installed Solar Panels Home	Bought Low-Energy Home	Electric Car
FR	87.7	68.0	54.5	40.4	29.9	24.4	19.6	14.8	11.1	8.8	2.4	7.5	1.2
BE	72.3	61.7	51.1	42.3	32.9	26.1	15.7	18.4	13.1	22.9	9.0	9.5	3.3
NL	87.8	84.8	77.6	67.1	39.6	39.6	50.3	20.7	30.2	24.5	21.5	12.7	2.8
DE	82.3	77.4	60.1	55.9	18.3	24.8	12.2	13.4	20.9	20.5	7.0	3.4	0.5
IT	66.5	41.7	40.2	21.3	14.1	4.8	7.5	9.4	4.2	10.9	5.5	3.2	2.3
LU	87.5	78.2	60.2	45.9	33.1	31.7	22.7	18.9	20.6	13.1	6.1	12.5	4.9
DK	79.8	72.8	65.6	51.7	36.0	31.6	27.4	27.0	21.4	17.3	6.6	10.6	1.6
IE	73.5	60.3	36.6	39.9	30.4	25.0	19.6	14.7	11.8	14.1	6.9	8.8	2.6
UK	83.4	65.9	42.0	48.9	33.9	26.7	28.5	16.6	17.1	21.6	4.0	5.4	2.9
GR	75.7	65.2	43.4	31.2	25.7	2.1	2.7	4.6	2.8	3.0	16.9	1.7	0.5
ES	85.5	62.1	39.7	33.7	17.6	10.6	17.6	4.4	3.6	4.4	1.6	2.8	0.8
PT	72.8	54.0	39.1	17.1	17.1	6.8	6.2	4.4	4.2	5.2	4.0	2.0	1.0
FI	81.5	76.2	48.4	45.8	20.9	41.3	21.3	16.6	26.0	16.1	3.4	3.4	1.9
SE	91.8	84.0	59.7	67.5	16.0	41.3	15.5	20.1	40.7	26.8	3.5	2.1	2.6
AT	61.5	58.5	46.2	41.5	17.6	23.8	19.1	12.3	18.3	15.1	6.2	3.4	3.4
CY	84.1	73.8	56.5	22.1	32.9	10.9	7.9	16.2	5.0	3.2	24.4	5.9	1.5
CZ	76.1	54.7	53.5	29.2	20.2	8.3	8.9	11.0	5.6	4.8	4.8	1.9	1.3
EE	75.3	64.5	50.8	32.4	35.6	11.4	11.4	13.2	3.8	2.9	1.8	4.9	0.6
HU	59.2	56.3	46.8	20.9	24.3	13.3	9.3	9.0	10.9	5.3	5.5	6.7	2.0
LV	63.1	62.1	64.9	40.8	30.6	11.8	15.7	14.8	4.3	4.8	1.2	1.5	1.2
LT	79.1	51.3	51.2	22.1	17.3	6.2	8.0	6.8	3.6	0.2	0.6	1.1	0.3
MT	84.9	59.9	48.0	30.9	5.6	13.5	28.3	5.9	6.6	1.0	13.8	2.3	1.6
PL	58.3	50.0	39.4	19.8	11.9	4.0	8.3	4.7	2.9	3.2	2.3	2.9	0.5
SK	71.3	54.8	36.5	28.0	25.6	5.5	9.3	8.7	2.8	4.3	3.1	0.7	1.0
SI	86.1	75.2	64.5	48.0	29.9	26.7	16.0	17.4	9.3	16.5	4.4	7.4	1.2
BG	34.1	37.7	53.2	24.8	36.8	2.4	3.3	5.7	1.9	1.7	3.7	2.7	0.5
RO	35.7	37.9	38.6	22.9	25.9	4.3	6.7	5.2	4.0	3.1	2.6	1.2	1.2
HR	66.3	47.8	40.0	21.2	23.7	5.5	5.8	8.2	3.5	4.4	0.8	2.6	0.1
EU28	73.4	62.0	50.3	37.1	25.0	17.7	14.7	12.4	11.8	10.7	5.7	4.5	1.6

Table A3. Multilevel mixed-effects logistic regression of climate change mitigation actions among European Union (EU) citizens.

	Reduce And Separate Waste		Use Fewer Disposable Items		Energy Efficiency Households (Eff Hh) Appliances		Alternatives To Car Use		Insulated Home Better		Food Carbon Footprint		Energy Saving Equipment Home	
	β	exp(β)	β	exp(β) 1	β	exp(β)	β	exp(β)	β	exp(β)	β	exp(β)	β	exp(β)
Gender (Men)														
Women	0.23 ***	1.25	0.32 ***	1.38	0.14 ***	1.15	0.15 ***	1.16	−0.09 **	0.92	0.47 ***	1.6	−0.08	0.93
AGE (15–24 YEARS)														
25–34 Years	0.2 *	1.22	0.1	1.11	0.46 ***	1.57	−0.04	0.96	0.33 **	1.39	−0.1	0.9	0.38 **	1.47
35–44 Years	0.34 ***	1.41	0.21 *	1.23	0.56 ***	1.75	−0.13	0.88	0.52 ***	1.69	−0.2	0.82	0.52 ***	1.68
45–54 Years	0.46 ***	1.59	0.2 *	1.22	0.62 ***	1.86	−0.17	0.85	0.68 ***	1.98	−0.21	0.81	0.46 ***	1.58
55–64 Years	0.45 ***	1.56	0.22 **	1.25	0.62 ***	1.85	−0.23 **	0.79	0.72 ***	2.05	−0.19	0.83	0.48 ***	1.62
65 Years and Over	0.39 ***	1.48	0.19 *	1.22	0.52 ***	1.68	−0.33 ***	0.72	0.82 ***	2.28	−0.27 *	0.77	0.48 ***	1.61
Age When Stopped Education (1–15)														
16–19	0.25 ***	1.28	0.15 **	1.16	0.32 ***	1.38	0.24 ***	1.27	0.24 ***	1.27	0.13	1.14	0.13	1.14
20+	0.46 ***	1.59	0.3 ***	1.35	0.5 ***	1.65	0.54 ***	1.72	0.4 ***	1.51	0.42 ***	1.52	0.26 ***	1.29
Still Studying	0.32 **	1.38	0.3 **	1.35	0.23 *	1.26	0.62 ***	1.86	0.27	1.31	0.33 **	1.4	0.19	1.21
No Full-Time Education	−0.03	0.97	−0.4	0.67	−0.09 *	0.92	0.46 ***	1.58	−0.38	0.69	0.58 *	1.79	−0.8 *	0.45
Occupation (Not Working)														
Self-Employed	−0.1	0.91	0.05	1.05	0.183 **	1.2	−0.18 **	0.84	0.37 ***	1.44	0.12	1.12	0.24 **	1.27
Employed	−0.13 **	0.88	−0.03	0.96	0.05	1.06	−0.13 **	0.89	0.005	1.01	0.04	1.04	−0.04	0.96
Marital Status (Single HH without Children)														
Single HH WITH Children	0.004	1.004	0.78	1.08	0.26 ***	1.3	0.01	1.01	0.17	1.18	0.1	1.1	0.12	1.13
Multiple HH without Children	0.05	1.06	0.07	1.07	0.27 ***	1.3	−0.12 **	0.89	0.49 ***	1.63	0.22 ***	1.24	0.35 ***	1.42
Multiple HH with Children	0.12 **	1.13	0.11 *	1.11	0.41 ***	1.51	0.08	0.92	0.63 ***	1.89	0.17 **	1.18	0.39 ***	1.47
Difficulties Paying Bills (Most of the Time														
From Time to Time	0.31 ***	1.37	0.26 ***	1.29	0.38 ***	1.47	−0.03	0.97	0.23 **	1.26	0.04	1.04	−0.04	0.96
Almost Never	0.62 ***	1.86	0.47 ***	1.6	0.58 ***	1.78	0.17 **	1.18	0.55 ***	1.74	0.21 *	1.23	0.12	1.13
Urban Area (Small Town)														
Large Town	0.001	1	−0.003	1	−0.08 **	0.93	0.21 ***	1.24	−0.22 ***	0.8	0.008	1.008	−0.22 ***	0.81
Region(East-Central Europe)														
Western Europe	0.67 **	1.95	0.63 ***	1.88	0.07	1.07	0.76 ***	2.15	−0.11	1.11	1.36 ***	3.89	0.96 ***	2.6
Southern Europe	0.68 **	1.97	0.24	1.2	−0.14	0.87	−0.05	0.95	−0.39 *	0.67	−0.06	0.94	0.13	1.14
Nordic Nations	0.67 *	1.95	0.31 **	2.21	−0.005	0.99	0.86 ***	2.36	−0.48 *	0.62	1.67 ***	5.27	0.76 **	2.15
Polit. Ideology (Right)														
Left	0.26 ***	1.3	0.2 ***	1.22	0.04	1.04	0.17 ***	1.18	−0.11 **	0.9	0.35 ***	1.42	−0.2 ***	0.82
Centre	0.21 ***	1.23	0.11 **	1.12	0.15 ***	1.16	0.06	1.06	0.07	1.07	0.05	1.05	−0.05	1.07
Political Interest Index (Not at All)														

Table A3. *Cont.*

	Reduce And Separate Waste		Use Fewer Disposable Items		Energy Efficiency Households (Eff Hh) Appliances		Alternatives To Car Use		Insulated Home Better		Food Carbon Footprint		Energy Saving Equipment Home	
	β	$exp(\beta)$	β	$exp(\beta)_1$	β	$exp(\beta)$	β	$exp(\beta)$	β	$exp(\beta)$	β	$exp(\beta)$	β	$exp(\beta)$
Strong	0.25 ***	1.29	0.6 ***	1.82	0.55 ***	1.74	0.63 ***	1.88	0.52 ***	1.69	0.88 ***	2.4	0.23 ***	1.26
Medium	0.34 ***	1.4	0.41 ***	1.51	0.39 ***	1.47	0.37 ***	1.45	0.32 ***	1.37	0.42 ***	1.52	−0.007	0.99
Low	0.33 ***	1.39	0.31 ***	1.37	0.3 ***	1.34	0.34 ***	1.41	0.16 *	1.17	0.17 *	1.18	−0.14	0.87
Climate Change Importance (Not a Serious Problem)														
A Fairly Serious Problem	0.17 *	1.19	0.21 **	1.23	0.1	1.1	0.13	1.14	0.64	1.26	0.16	1.17	−0.07	0.94
A Very Serious Problem	0.7 ***	2.02	0.81 ***	2.24	0.53 ***	1.7	0.65 ***	1.92	0.18 *	1.19	0.76 ***	2.13	0.0001	1
Climate Change Resp. (No)														
National Gov.	0.15 ***	1.16	0.18 ***	1.19	0.17 ***	1.18	0.09 **	1.09	0.07	1.07	−0.1 *	0.9	0.04	1.04
European Union	0.16 ***	1.17	0.79 *	1.08	0.14 ***	1.15	0.09 **	1.09	0.004	1.004	0.05	1.05	−0.03	0.97
Regional/Local Authorities	−0.00089 ***	1	0.86 *	1.09	0.77 *	1.08	−0.05	0.95	0.07	1.07	0.28 ***	1.18	0.13 **	1.14
Business and Industry	0.33 ***	1.4	0.17 ***	1.19	0.24 ***	1.27	0.04	1.04	0.06	1.06	0.05	1.05	−0.005	1
Citizens	0.21 ***	1.24	0.26 ***	1.3	0.16 ***	1.18	0.42 ***	1.53	0.14 ***	1.15	0.28 ***	1.32	0.17 ***	1.19
Env. Groups	−0.09 *	0.92	−0.002	1	−0.07	0.94	0.07	1.08	−0.07	0.93	0.05	1.05	0.13 **	1.14
Action on Climate Change will Lead to Innovation (No)														
Yes	0.28 ***	1.32	0.25 ***	1.29	0.25 ***	1.28	0.07	1.07	0.01	1.01	0.13	1.14	0.01	1.01
Constant	−2.04	0.13	−2.34	0.1	−2.95	0.05	−2.5	0.08	−3.17	0.04	−4.52	0.01	−3.07	0.05
Observations	13714		13714		13714		13714		13714		13714		13714	
No. of Groups	28		28		28		28		28		28		28	
Log Likelihood	−6726.7		−7989.75		8712.85		−8225.35		−7411.39		−5765.97		−5634.83	
Wald Chi [2]	673.63 ***		791.48 ***		891.36 ***		792.21 ***		607.97 ***		696.91 ***		245.83 ***	
Random Part Identity: Country														
Variance (Constant)	0.39		0.22		0.19		0.12		0.2		0.33		0.3	
(Intercept Variance) (Standard Error)	0.11		0.06		0. 05		0.04		0.06		0.1		0.09	
Variance at Country Level [2] (%)	10.59%		6.26%		4.53%		3.40%		5.73%		9.12%		8.36%	
LR Test	624.95 ***		344.68 ***		372.26 ***		248.94 ***		311.61 ***		246.27 ***		401.07 ***	

	Low Fuel Consumption Car		Travel Carbon Footprint		Switched Energy Supplier		Installed Solar panels Home		Bought Low-Energy Home		Electric car	
	β	$exp(\beta)$	β	$exp(\beta)$	β	$exp(\beta)$	β	$exp(\beta)$	β	$exp(\beta)$	β	$exp(\beta)$
Gender (Men)												
Women	−0.23 ***	0.8	0.2 ***	1.23	−0.003	1	−0.07	0.93	−0.17 **	0.84	−0.11	0.9
Age (15–24 Years)												
25–34 Years	−0.27 *	0.76	01	1.1	0.23	1.26	−0.006	0.99	0.12	1.13	−0.7 *	0.5
35–44 Years	−0.35 **	0.7	0.23	1.26	0.3 *	1.35	0.16	1.18	0.16	1.17	−0.62 *	0.54

Table A3. *Cont.*

	Low Fuel Consumption Car		Travel Carbon Footprint		Switched Energy Supplier		Installed Solar panels Home		Bought Low-Energy Home		Electric car	
	β	$exp(\beta)$	β	$exp(\beta)$	β	$exp(\beta)$	β	$exp(\beta)$	β	$exp(\beta)$	β	$exp(\beta)$
45–54 Years	−0.005	0.99	0.14	1.15	0.24	1.27	0.23	1.26	−0.1	0.9	−0.34	0.71
55–64 Years	−0.16	0.85	0.07	1.07	0.26	1.29	0.5	1.64	−0.08	0.92	−0.32	0.73
65 Years and Over	−0.27	0.77	0.02	1.02	0.18	1.2	0.33 **	1.38	−0.08	0.93	−0.74 *	0.48
Age When Stopped Education (1–15)												
16–19	0.17	1.19	0.05	1.05	0.06	1.06	0.36 **	1.43	0.06	1.06	0.14	1.15
20+	0.39 ***	1.47	0.46 ***	1.59	0.31 ***	1.37	0.69 ***	1.98	0.31 *	1.36	0.48	1.61
Still Studying	−0.32	0.72	0.55 ***	1.73	0.05	1.05	0.7 **	2	0.45	1.56	0.26	1.3
No Full-Time Education	−0.52	0.6	0.57	1.78	0.7 *	2.01	0.3	1.35	−0.42	0.66	0.52	1.67
Occupation (Not Working)												
Self-Employed	0.31 ***	1.36	0.46	1.58	0.18 *	1.2	0.3 **	1.35	0.32 *	1.38	0.97 ***	2.63
Employed	0.02	1.02	0.02 ***	1.02	−0.11	0.89	−0.13	0.88	0.1	1.11	0.38 *	1.46
Marital Status (Single HH without Children)												
Single HH with Children	0.25 *	1.29	−0.04	0.97	0.28 **	1.32	0.22	1.25	0.36 *	1.43	−0.24	0.78
Multiple HH Without Children	0.6 ***	1.82	0.08	1.08	0.3 ***	1.35	0.37 ***	1.45	0.31 ***	1.36	0.25	1.28
Multiple HH with Children	0.71 ***	2.04	0.08	1.08	0.49 ***	1.63	0.62 ***	1.86	0.56 ***	1.75	0.56 ***	1.75
Difficulties Paying Bills (Most of the Time)												
From Time to Time	0.19	1.21	0.34 **	1.4	−0.16	0.85	−0.1	0.91	−0.06	0.94	−0.12	0.89
Almost Never	0.67 ***	1.95	0.36 **	1.43	−0.02	0.97	0.17	1.18	0.36 *	1.44	0.23	1.26
Urban Area (Small Town)												
Large Town	−0.12 **	0.89	0.14 ***	1.15	0.05	1.05	−0.27 ***	0.76	−0.21 **	0.81	0.07	1.07
Region (East-Central Europe)												
Western Europe	0.53 ***	1.71	1.33 ***	3.78	1.53 ***	4.61	0.9 ***	2.46	0.94 ***	2.57	0.94 ***	2.55
Southern Europe	−0.2	0.82	−0.1	0.91	0.12	1.13	1.4 ***	4.02	0.21	1.23	0.37	1.45
Nordic Nations	0.69 ***	1.98	1.84 ***	6.32	1.56 ***	4.74	0.16	1.18	0.28	1.32	0.6 *	1.82
Polit. Ideology (Right)												
Left	−0.22 ***	0.8	0.59 ***	1.81	0.09	1.09	−0.09	0.92	−0.09	0.92	−0.39 **	0.68
Center	−0.09	0.91	0.34 ***	1.41	0.01	1.02	−0.16	0.85	0.07	1.07	−0.38 *	0.68
Political Interest Index (Not at All)												
Strong	0.44 ***	1.55	0.76	2.14	0.66 ***	1.93	0.19	1.21	0.32 **	1.37	−0.1	0.91
Medium	0.25 ***	1.28	0.27	1.3	0.37 ***	1.45	0.01	1.01	0.05	1.05	−0.31	0.73
Low	0.17	1.19	0.11	1.12	0.32 ***	1.38	−0.06	0.94	−0.06	0.94	−0.13	0.88

Table A3. *Cont.*

	Low Fuel Consumption Car		Travel Carbon Footprint		Switched Energy Supplier		Installed Solar panels Home		Bought Low-Energy Home		Electric car	
	β	$exp(\beta)$	β	$exp(\beta)$	β	$exp(\beta)$	β	$exp(\beta)$	β	$exp(\beta)$	β	$exp(\beta)$
Climate Change Importance (not a Serious Problem)												
A Fairly Serious Problem	0.12	1.13	0.5	1.64	−0.01	0.99	−0.17	0.84	0.21	1.23	−0.12	0.89
A very Serious Problem	0.3 **	1.35	0.93	2.54	0.26 **	1.3	−0.06	0.95	0.2	1.22	0.07	1.07
Climate Change Resp. (No)												
National Gov.	−0.06	0.94	−0.07	0.93	0.09	1.1	0.13	1.14	0.06	1.06	−0.14	0.87
European Union	0.06	1.06	0.05	1.05	0.05	0.05	0.09	1.1	−0.04	0.96	−0.04	0.97
Regional/Local Authorities	−0.04	0.96	0.009	1.01	0.05	1.05	0.15	1.16	0.07	1.07	0.36 **	1.44
Business and Industry	0.02	1.02	−0.14	0.99	−0.03	0.97	−0.17	0.85	−0.21 **	0.81	0.006	1.01
Citizens	0.19 ***	1.21	0.38 ***	0.99	−0.18 ***	1.2	0.22	1.25	0.11	1.12	0.07	1.07
Env. Groups	0.06	1.06	0.1	1.1	0.1	1.1	0.38	1.04	0.3	1.03	0.06	1.07
Action on Climate Change will Lead to Innovation (no)												
Yes	−0.05	0.95	0.09	1.09	0.22 **	1.25	0.03	1.03	0.05	1.05	−0.19	0.83
Constant	−3.52	0.03	−5.61	0.004	−4.68	0.01	−4.64	0.01	−4.43	0.01	−4.62	0.01
Observations	13714		13714		13714		13714		13714		13714	
No. of Groups	28		28		28		28		28		28	
Log likelihood	−5127.18		−4582.28		−4585.54		−2917.55		−2576.41		−1144.6	
Wald Chi [2]	424.58 ***		572.44 ***		279.46 ***		190.72 ***		135 ***		107.2 ***	
Random Part Identity: Country												
Variance (Constant)	0.14		0.16		0.34		0.46		0.27		0.16	
(Intercept Variance) (Standard Error)	0.05		0.05		0.11		0.14		0.09		0.08	
Variance at Country Level [1] (%)	4.08%		4.64%		9.37%		12.27%		7.58%		4.64%	
LR Test	87.13 ***		147.19 ***		180.31		213.54		111.08		12.8	

Note: All coefficients are compared to the benchmark category, shown in brackets. All country level indicators were centred to the sample mean. For the sample of 18,529 respondents. *** $p < 0.01$, ** $p < 0.05$, * $p < 0.1$. [1] Odds ratio. [2] Variance partition coefficient: measures the proportion of the total residual variance that is due to between-group variation.

Table A4. Multilevel mixed-effects logistic regression of climate change mitigation cumulative actions in EU.

	Model I		Model II		Model III		Model IV		Model V		Model VI	
	β	exp(β) [1]	β	exp(β)	β	exp(β)	β	exp(β)	β	exp(β)	β	exp(β)
Gender (men)												
Women	0.21 ***	1.23	0.21 ***	1.23	0.21 ***	1.23	0.21 ***	1.23	0.21 ***	1.23	0.21 ***	1.23
Age (15–24 years)												
25–34 years	0.36 ***	1.43	0.35 ***	1.42	0.35 ***	1.43	0.35 ***	1.43	0.35 ***	1.43	0.36 ***	1.42
35–44 years	0.37 ***	1.45	0.37 ***	1.45	0.37 ***	1.45	0.37 ***	1.45	0.37 ***	1.45	0.37 ***	1.45
45–54 years	0.46 ***	1.58	0.46 ***	1.58	0.46 ***	1.58	0.46 ***	1.58	0.46 ***	1.58	0.46 ***	1.58
55–64 years	0.49 ***	1.63	0.49 ***	1.63	0.5 ***	1.63	0.49 ***	1.63	0.49 ***	1.63	0.49 ***	1.64
65 years and over	0.45 ***	1.57	0.45 ***	1.57	0.45 ***	1.57	0.45 ***	1.57	0.45 ***	1.57	0.45 ***	1.57
Age when stopped Education (1–15)												
16–19	0.31 ***	1.36	0.31 ***	1.36	0.31 ***	1.36	0.31 ***	1.36	0.31 ***	1.36	0.31 ***	1.36
20+	0.7 ***	2.02	0.7 ***	2.02	0.7 ***	2.02	0.7 ***	2.02	0.7 ***	2.03	0.7 ***	2.02
Still studying	0.49 ***	1.64	0.49 ***	1.63	0.49 ***	1.63	0.49 ***	1.63	0.49 ***	1.64	0.5 ***	1.64
No full-time education	0.12	1.12	0.12	1.13	0.12	1.12	0.12	1.12	0.12	1.12	0.12	1.12
Occupation (not working)												
Self-employed	0.29 ***	1.34	0.29 ***	1.34	0.29 ***	1.34	0.29 ***	1.34	0.3 ***	1.34	0.29 ***	1.35
Employed	−0.07	0.93	−0.07	0.93	−0.07	0.93	−0.07	0.93	−0.07	0.93	−0.07	0.94
Marital status (single HH without children)												
Single HH with children	0.28 ***	1.33	0.28 ***	1.33	0.28 ***	1.33	0.28 ***	1.33	0.28 ***	1.33	0.28 ***	1.33
Multiple HH without children	0.35 ***	1.42	0.35 ***	1.42	0.35 ***	1.42	0.35 ***	1.42	0.35 ***	1.42	0.35 ***	1.42
Multiple HH with children	0.53 ***	1.7	0.53 ***	1.7	0.53 ***	1.7	0.53 ***	1.7	0.53 ***	1.7	0.53 ***	1.7
Difficulties paying bills (most of the time												
From time to time	0.3 ***	1.34	0.3 ***	1.34	0.3 ***	1.35	0.3 ***	1.35	0.3 ***	1.34	0.3 ***	1.34
Almost never	0.7 **	2.01	0.7 ***	2.01	0.7 ***	2.02	0.7 ***	2.01	0.7 ***	2	0.7 **	1.99
Urban area (small town)												
Large town	−0.11 **	0.89	−0.11 **	0.89	−0.11 **	0.89	−0.11 **	0.89	−0.11	0.89	−0.11 **	0.89
Region (East-Central Europe)												
Western Europe	1.14 ***	3.11	0.69 *	2	0.71 *	3.95	0.71 *	2.03	0.87 ***	2.38	0.72 ***	2.05
Southern Europe	0.26	1.3	0.13	1.14	−0.07	1.42	−0.07	0.94	0.58 **	1.78	0.45 **	1.57
Nordic nations	1.21 ***	3.35	0.57	1.76	0.55	4.13	0.55	1.73	0.5	1.65	0.38	1.47
Polit. ideology (right)												
Left	0.21 ***	1.23	0.21 ***	1.23	0.21 ***	1.24	0.21 ***	1.23	0.21 ***	1.23	0.21 ***	1.23
Centre	0.1 *	1.11	1 *	1.1	0.1 *	1.11	0.1 *	1.11	0.1 *	1.11	0.1 *	1.11
Political interest index (not at all)												
Strong	0.9 ***	2.45	0.9 ***	2.45	0.9 ***	2.45	0.9 ***	2.46	0.9 ***	2.45	0.9 ***	2.45
Medium	0.46 ***	1.59	0.46 ***	1.59	0.47 ***	1.59	0.47 ***	1.59	0.46 ***	1.59	0.47 ***	1.59
Low	0.32 ***	1.38	0.32 ***	1.38	0.32 ***	1.38	0.32 ***	1.38	0.32 ***	1.38	0.32 ***	1.38

Table A4. *Cont.*

	Model I		Model II		Model III		Model IV		Model V		Model VI	
	β	$exp(\beta)$ [1]	β	$exp(\beta)$	β	$exp(\beta)$	β	$exp(\beta)$	β	$exp(\beta)$	β	$exp(\beta)$
Climate change												
Importance (not a serious problem)												
A fairly serious problem	0.21 *	1.23	0.21 *	1.23	0.21 *	1.23	0.21 *	1.23	0.21 *	1.23	0.21 *	1.23
A very serious problem	0.86 ***	2.37	0.86 ***	2.37	0.86 ***	2.37	0.86 ***	2.37	0.86 ***	2.37	0.86 ***	2.37
Climate change resp. (no)												
National gov.	0.13 **	1.13	0.12 **	1.13	0.13 **	1.13	0.13 **	1.13	0.12 **	1.13	0.12 **	1.13
European Union	0.04	1.04	0.04	1.04	0.04	1.04	0.04	1.04	0.04	1.04	0.04	1.04
Regional/local authorities	0.11 **	1.12	0.11 **	1.12	0.11 **	1.12	0.11 **	1.12	0.11 **	1.12	0.11 **	1.12
Business and industry	0.16 ***	1.17	0.16 ***	1.17	0.16 ***	1.17	0.16 ***	1.17	0.16 ***	1.17	0.16 ***	1.17
Citizens	0.38 ***	1.46	0.38 ***	1.46	0.38	1.46	0.38 ***	1.46	0.38 ***	1.46	0.38 ***	1.46
Env. groups	−0.07	0.93	−0.07	0.93	−0.07	0.93	−0.07	0.93	−0.07	0.93	−0.07	0.93
Action on climate change will lead to innovation (no)												
Yes	0.16 **	1.18	0.16 **	1.18	0.16 **	1.18	0.16 **	1.18	0.16	1.18	0.16 **	1.18
GDP/capita	$8.43 * 10^{-6}$	1										
HDI			10.27 **	28,935.8								
HPI					−0.014	0.99						
SPI							0.09 **	1.09				
Internet skills									0.04 **	1.04		
Internet access											0.82 ***	1.09
Government effectiveness index												
Policy performance index												
Environment												
Large town *HDI												
Large town *SPI												
Large town *environment												
Constant	−5.2	0.006	−4.96	0.007	−4.92	0.005	−4.92	0.5	−5.09	0.006	−11.77	$7.7 * 10^{-7}$
Observations	13,714		13,714		13,714		13,714		13,714		13,714	
No. of groups	28		28		28		28		28		28	
Log likelihood	−6803.1		−6801.51		−6803.15		−6801.55		−6800.54		−6798.35	
Wald Chi [2]	1030.58 ***		1038.71		1030.28		1038.65		1043.86		1057.57 ***	
RANDOM PART IDENTITY: COUNTRY												
Variance (constant)	0.28		0.25		0.28		0.25		0.23		0.19	
(Intercept variance) (standard error)	0.08		0.07		0.08		0.07		0.07		0.06	
Variance at country level [2] **(%)**	7.84%		7.06%		7.84%		7.06%		6.53%		5.46%	
LR test	369.54 ***		338.71 ***		370.31 ***		309.61 ***		259.76 ***		229.46 ***	
Gender (men)												
Women	0.21 ***	1.23	0.21 ***	1.23	0.21 ***	1.23	0.2 ***	1.23	0.2 ***	1.23	0.2 ***	1.23

Table A4. *Cont.*

	Model I		Model II		Model III		Model IV		Model V		Model VI	
	β	$exp(\beta)$ [1]	β	$exp(\beta)$	β	$exp(\beta)$	β	$exp(\beta)$	β	$exp(\beta)$	β	$exp(\beta)$
Age (15–24 years)												
25–34 years	0.36 ***	1.43	0.35 ***	1.43	0.35 ***	1.43	0.35 ***	1.43	0.35 ***	1.42	0.36 ***	1.43
35–44 years	0.37 ***	1.45	0.37 ***	1.45	0.37 ***	1.45	0.37 ***	1.45	0.37 ***	1.45	0.38 ***	1.46
45–54 years	0.46 ***	1.58	0.46 ***	1.58	0.46 ***	1.58	0.46 ***	1.58	0.46 ***	1.58	0.46 ***	1.58
55–64 years	0.49 ***	1.63	0.49 ***	1.63	0.49 ***	1.63	0.49 ***	1.63	0.49 ***	1.63	0.5 ***	1.64
65 years and over	0.45 ***	1.57	0.45 ***	1.57	0.45 ***	1.57	0.45 ***	1.57	0.45 ***	1.57	0.45 ***	1.57
Age when stopped Education (1–15)												
16–19	0.31 ***	1.36	0.31 ***	1.36	0.31 ***	1.36	0.31 ***	1.36	0.31 ***	1.36	0.31 ***	1.37
20+	0.7 ***	2.02	0.7 ***	2.02	0.7 ***	2.02	0.7 ***	2.02	0.7 ***	2.01	0.7 ***	2.02
Still studying	0.5 ***	1.63	0.49 ***	1.64	0.49 ***	1.64	0.49 ***	1.63	0.49 ***	1.63	0.49 ***	1.64
No full-time education	0.12	1.13	0.12	1.12	0.12	1.12	0.12	1.13	0.12	1.13	0.12	1.13
Occupation (not working)												
Self-employed	0.29 ***	1.34	0.29 ***	1.34	0.29 ***	1.34	0.29 ***	1.34	0.29 ***	1.33	0.29 ***	1.33
Employed	−0.07	0.93	−0.07	0.93	−0.07	0.93	−0.07	0.93	−0.07	0.93	−0.07	0.93
Marital status (single HH without children)												
Single HH with children	0.28 ***	1.33	0.28 ***	1.33	0.28 ***	1.33	0.28 ***	1.33	0.28 ***	1.33	0.28 ***	1.33
Multiple HH without children	0.35 ***	1.42	0.34 ***	1.42	0.35 ***	1.42	0.35 ***	1.42	0.35 ***	1.42	0.35 ***	1.42
Multiple HH with children	0.53 ***	1.7	0.53 ***	1.7	0.53 ***	1.7	0.53 ***	1.7	0.53 ***	1.7	0.53 ***	1.7
Difficulties paying bills (most of the time)												
From time to time	0.3 ***	1.34	0.29 ***	1.34	0.3 ***	1.34	0.3 ***	1.35	0.3 ***	1.35	0.3 ***	1.35
Almost never)	0.7 **	2.02	0.7 **	2	0.7 **	2.02	0.7 **	2.02	0.7 **	2.02	0.7 **	2.02
Urban area (small town)												
Large town	−0.11 **	0.89	−0.11 **	0.89	−0.11 **	0.89	−0.14 **	0.87	−0.14 **	0.87	−0.13 **	0.88
Region (East-Central Europe)												
Western Europe	1.38 ***	4	0.91 ***	2.49	1.25 ***	3.49	0.69 *	2	0.71 *	2.04	1.25 ***	3.5 ***
Southern Europe	0.28	1.33	0.43 *	1.54	0.51 *	1.67	0.13	1.13	−0.06	0.94	0.52 *	1.68 *
Nordic nations	1.46 **	4.3	0.51 **	1.67	0.96 **	2.61	0.57	1.77	0.55	1.75	0.97 **	2.65 **
Polit. ideology (right)												
Left	0.21 ***	1.23	0.21 ***	1.23	0.21 ***	1.23	0.21 ***	1.23	0.21 ***	1.23	0.21 ***	1.24
Centre	0.1 *	1.11	0.1 *	1.1	0.1 *	1.1	0.1 *	1.1	0.1 *	1.1	0.1 *	1.11
Political interest index (not at all)												
Strong	0.9 ***	2.45	0.9 ***	2.46	0.9 ***	2.45	0.89 ***	2.45	0.9 ***	2.45	0.9 ***	2.45
Medium	0.46 ***	1.59	0.47 ***	1.59	0.46 ***	1.59	0.46 ***	1.59	0.46 ***	1.59	0.46 ***	1.59
Low	0.32 ***	1.38	0.32 ***	1.38	0.32 ***	1.38	0.32 ***	1.38	0.32 ***	1.38	0.32 ***	1.38
Climate change Importance (not a serious problem)												
A fairly serious problem	0.21 *	1.23	0.21 *	1.23	0.21 *	1.23	0.2 *	1.23	0.2 *	1.23	0.21 *	1.23
A very serious problem	0.86 ***	2.37	0.86 ***	2.36	0.86 ***	2.37	0.85 ***	2.36	0.85 ***	2.36	0.86 ***	2.37

Table A4. *Cont.*

	Model I		Model II		Model III		Model IV		Model V		Model VI	
	β	$exp(\beta)$ [1]	β	$exp(\beta)$	β	$exp(\beta)$	β	$exp(\beta)$	β	$exp(\beta)$	β	$exp(\beta)$
Climate change resp. (no)												
National gov.	0.13 **	1.13	0.12 **	1.13	0.13 **	1.13	0.13 **	1.13	0.13 **	1.13	0.13 **	1.13
European Union	0.04	1.04	0.04	1.04	0.04	1.04	0.04	1.04	0.04	1.04	0.04	1.04
Regional/local authorities	0.11 **	1.12	0.11 **	1.12	0.11 **	1.12	0.11 **	1.12	0.11 **	1.12	0.11 **	1.12
Business and industry	0.16 ***	1.17	0.16 ***	1.17	0.16 ***	1.17	0.16 ***	1.17	0.16 ***	1.16	0.16 ***	1.16
Citizens	0.38 ***	1.46	0.38 ***	1.46	0.38 ***	1.46	0.38 ***	1.47	0.38 ***	1.47	0.38 ***	1.46
Env. groups	−0.07	0.93	−0.07	0.93	−0.07	0.93	−0.07	0.94	−0.07	0.94	−0.07	0.93
Action on climate change will lead to innovation (no)												
Yes	0.16 **	1.18	0.16 **	1.18	0.16 **	1.18	0.17	1.18	0.17	1.18	0.16	1.18
GDP/capita												
HDI							9.31 *	11040.3				
HPI												
SPI									0.08 *	1.08		
Internet skills												
Internet access												
Government effectiveness index	−0.05	0.95										
Policy performance index			0.43 **	1.54								
Environment					0.23 *	1.26					0.2	1.22
Large town *HDI							2.16 *	8.67				
Large town *SPI									0.02 *	1.02		
Large town *environment											0.08 *	1.08
Constant	−5.09	0.006	−5.07	0.006	−5.24	−0.005	−4.95	−0.007	−4.91	0.007	−5.24	0.005
Observations	13,714		13,714		13,714		13,714		13,714		13,714	
No. Of groups	28		28		28		28		28		28	
Log likelihood	−6803.27		−6800.65		−6801.78		−6801.05		−6799.86		−6800.17	
Wald Chi [2]	1029.72 ***		1043.3 ***		1037.06 ***		1040.85 ***		1041.26 ***		1039.57 ***	
Random part identity: country												
Variance (constant)	0.29		0.23		0.26		0.25		0.25		0.25	
(Intercept variance) (standard error)	0.08		0.07		0.07		0.07		0.07		0.25	
Variance at country level (%)	8.1%		6.53%		7.32%		7.06%		7.06%		7.06%	
LR test	369.51 ***		296.61 ***		338.27 ***		340.07 ***		310.78 ***		337.01 ***	

[1] Odds ratio. [2] Variance partition coefficient: measures the proportion of the total residual variance that is due to between-group variation. *** $p < 0.01$, ** $p < 0.05$, * $p < 0.1$.

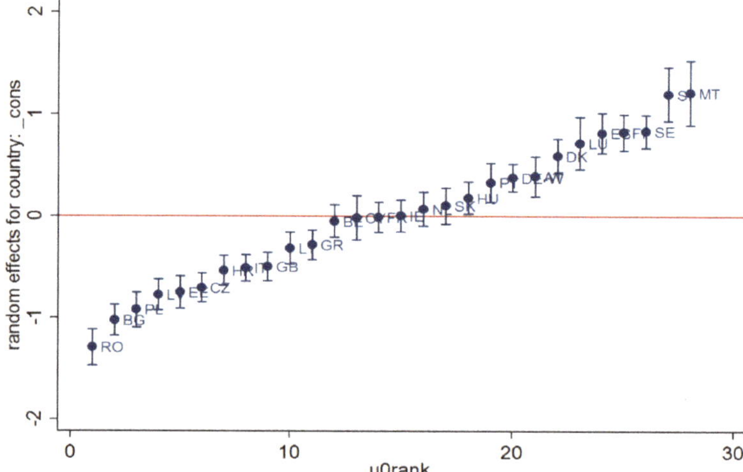

Figure A1. Country-level effects of null model for the decision to undertake climate change individual actions. **Source:** Author's contribution.

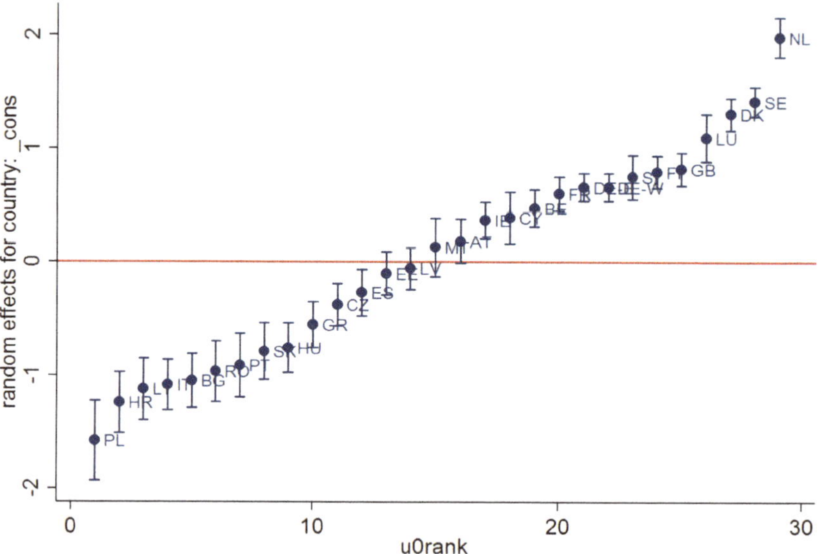

Figure A2. Country-level effects of null model of extra mitigation behaviour. **Source:** Author's contribution.

References

1. European Investment Bank. The EIB Circular Economy Guide. Supporting the Circular Transition. Available online: https://www.eib.org/attachments/thematic/circular_economy_guide_en.pdf (accessed on 8 April 2020).
2. Whitmarsh, L. Scepticism and uncertainty about climate change: Dimensions, determinants and change over time. *Glob. Environ. Chang.* **2011**, *21*, 690–700. [CrossRef]
3. Whitmarsh, L.; Seyfang, G.; O'Neill, S. Public engagement with carbon and climate change: To what extent is the public 'carbon capable'? *Glob. Environ. Chang.* **2011**, *21*, 56–65. [CrossRef]

4. Thøgersen, J.; Crompton, T. Simple and painless? The limitations of spillover in environmental campaigning. *J. Consum. Pol.* **2009**, *32*, 141–163. [CrossRef]
5. Ortega-Egea, J.M.; García-de-Frutos, N.; Antolín-López, R. Why do some people do "more" to mitigate climate change than others? Exploring heterogeneity in psycho-social associations. *PLoS ONE* **2014**, *9*, e106645. [CrossRef]
6. Zamfir, A.M.; Mocanu, C.; Grigorescu, A. Circular Economy and decision models among European SMEs. *Sustainability* **2017**, *9*, 1507. [CrossRef]
7. GESIS Data Archive for the Social Sciences. Available online: https://www.gesis.org/index.php?id=12376 (accessed on 8 May 2020).
8. Breure, A.M.; Lijzen, J.P.A.; Maring, L. Soil and land management in a circular economy. *Sci. Total Environ.* **2018**, *624*, 1125–1130. [CrossRef]
9. Shao, W.C.; Dong, Y.W. Study on the feasibility of improving the indoor air quality from the perspective of circular economy by activation of biological calcium. In Proceedings of the 2019 IEEE International Conference on Architecture, Construction, Environment and Hydraulics (ICACEH), Xiamen, China, 20–22 December 2019; pp. 133–136.
10. Blomsma, F.; Brennan, G. The emergence of circular economy: A new framing around prolonging resource productivity. *J. Ind. Ecol.* **2017**, *21*, 603–614. [CrossRef]
11. Naustdalslid, J. Circular economy in China—The environmental dimension of the harmonious society. *Int. J. Sustain. Dev. World Ecol.* **2017**, *21*, 303–313. [CrossRef]
12. Ma, S.; Hu, S.; Chen, D.; Zhu, B. A case study of a phosphorus chemical firm's application of resource efficiency and eco-efficiency in industrial metabolism under circular economy. *J. Clean. Prod.* **2015**, *87*, 839–849. [CrossRef]
13. Korhonen, J.; Honkasalo, A.; Seppala, J. Circular Economy: The Concept and its Limitations. *Ecol. Econ.* **2018**, *143*, 37–46. [CrossRef]
14. Haas, W.; Krausmann, F.; Wiedenhofer, D.Y.; Heinz, M. How circular is the global economy? An assessment of material flows, waste production, and recycling in the European Union and the world in 2005. *J. Ind. Ecol.* **2015**, *19*, 765–777. [CrossRef]
15. Wu, H.; Shi, Y.; Xia, Q.; Zhu, W. Effectiveness of the policy of circular economy in China: A DEA-based analysis for the period of 11th five-year-plan. *Resour. Conserv. Recycl.* **2014**, *83*, 163–175. [CrossRef]
16. Esposito, M.; Tse, T.; Soufani, K. Is the circular economy a new fast-expanding market? *Thunderbird Int. Bus. Rev.* **2015**, *49*, 630–631. [CrossRef]
17. Castellani, V.; Sala, S.; Mirabella, N. Beyond the throwaway society: A life cycle-based assessment of the environmental benefit of reuse. *Integr. Environ. Assess. Manag.* **2015**, *11*, 373–382. [CrossRef]
18. Keesstra, S.D.; Bouma, J.; Wallinga, J.; Tittonell, P.; Smith, P.; Cerdà, A.; Montanarella, L.; Quinton, J.N.; Pachepsky, Y.; Van der Putten, W.H.; et al. The significance of soils and soil science towards realization of the United Nations Sustainable Development Goals. *Soil* **2016**, *2*, 111–128. [CrossRef]
19. Ellen MacArthur Foundation. What Is a Circular Economy? Available online: https://www.ellenmacarthurfoundation.org/circular-economy/concept (accessed on 12 April 2020).
20. Schut, E.; Crielaard, M.; Mesman, M. Economy in the Dutch Construction Sector: A Perspective for the Market and Government. Available online: http://www.rivm.nl/dsresource?objectid=806b288e-3ae9-47f1-a28f-7c208f884b36&type=org&disposition=inline (accessed on 11 June 2020).
21. Geissdoerfer, M.; Savaget, P.; Bocken, N.M.P.; Hultink, E.J. The circular economy – a new sustainability paradigm. *J. Clean. Prod.* **2017**, *22*, 757–768. [CrossRef]
22. Open Data and the Circular Economy. Available online: https://www.europeandataportal.eu/en/highlights/open-data-and-circular-economy (accessed on 12 April 2020).
23. Van Buren, N.; Demmers, M.; Van der Heijden, R.; Witlox, F. Towards a Circular Economy: The Role of Dutch Logistics Industries and Governments. *Sustainability* **2016**, *8*, 647. [CrossRef]
24. The Circular Economy Foundation. Circular Economy. Towards and Eco-Efficient Europe. Available online: http://economiacircular.org/EN/?page_id=62 (accessed on 21 April 2020).
25. UN-HABITAT. Global Report on Human Settlements 2011: Cities and Climate Change. Available online: https://unhabitat.org/global-report-on-human-settlements-2011-cities-and-climate-change (accessed on 18 April 2020).

26. Camaren, P.; Swilling, M. Sustainable Resource Efficient Cities: Making it Happen. Nairobi: United Nations Environment Programme. Available online: https://www.academia.edu/12776947/Sustainable_Resource_Efficient_Cities_Making_it_Happen (accessed on 15 April 2020).
27. Leipold, S.; Petit-Boix, A. The circular economy and the bio-based sector - perspectives of European and German stakeholders. *J. Clean. Prod.* **2018**, *201*, 1125–1137. [CrossRef]
28. Liang, S.; Zhang, T. Data acquisition for applying input-output tables in Chinese cities. *J. Ind. Ecol.* **2011**, *15*, 825–835. [CrossRef]
29. Mihai, M.; Manea, D.; Titan, E.; Vasile, V. Correlations in the European Circular Economy. *Econ. Comput. Econ. Cybern. Stud. Res.* **2018**, *52*, 61–78.
30. Vercalsteren, A.; Christis, M.; Van Hoof, V. Indicators for a Circular Economy. Available online: https://vlaanderen-circulair.be/en/summa-ce-centre/publications/indicators-for-a-circular-economy (accessed on 10 April 2020).
31. Circular Economy Could Prevent Climate Change. Available online: https://www.therecycler.com/posts/circular-economy-could-prevent-climate-change (accessed on 12 May 2020).
32. Rutherford, O. Transition to Circular Economy must Be Part of Climate Strategy, Says CCC. Available online: https://resource.co/article/transition-circular-economy-must-be-part-climate-strategy-says-ccc (accessed on 22 May 2020).
33. Intergovernmental Panel on Climate Change (IPCC). AR4 Climate Change 2007: Synthesis Report. Available online: https://www.ipcc.ch/report/ar4/syr (accessed on 17 May 2020).
34. Riahi, K.; Grübler, A.; Nakicenovic, N. Scenarios of long-term socio-economic and environmental development under climate stabilization. *Technol. Forecast. Soc. Chang.* **2007**, *74*, 887–935. [CrossRef]
35. Luderer, G.; Pietzcker, R.C.; Bertram, C.; Kriegler, E.; Meinshausen, M.; Edenhofer, O. Economic mitigation challenges: How further delay closes the door for achieving climate targets. *Environ. Res. Lett.* **2013**, *8*, 1–8. [CrossRef]
36. Lorenzoni, I.; Nicholson-Cole, S.; Whitmarsh, L. Barriers perceived to engaging with climate change among the UK public and their policy implications. *Glob. Environ. Chang.* **2007**, *17*, 445–459. [CrossRef]
37. Pielke, R.A. Misdefining "climate change": Consequences for science and action. *Environ. Sci. Policy* **2005**, *8*, 548–561. [CrossRef]
38. IPCC. *IPCC Second Assessment Report: Climate Change*; Cambridge University Press: Cambridge, UK, 1995.
39. Karl, T.R.; Trenberth, K.E. Modern Global Climate Change. *Science* **2003**, *302*, 1719–1723. [CrossRef] [PubMed]
40. McCarthy, J.J.; Canziani, O.F.; Leary, N.A.; Dokken, D.J.; White, K.S. *Climate Change 2001: Impacts, Sdaptation, and Vulnerability*; Cambridge University Press: Cambridge, UK, 2001.
41. Semenza, J.C.; Hall, D.E.; Wilson, D.J.; Bontempo, B.D.; Sailor, D.J.; George, L.A. Public perception of climate change voluntary mitigation and barriers to behavior change. *Am. J. Prev. Med.* **2008**, *35*, 479–487. [CrossRef]
42. Betsill, M.M. Mitigating Climate Change in US Cities: Opportunities and obstacles. *Local Environ.* **2001**, *6*, 393–406. [CrossRef]
43. Clarke, J.; Newman, J.; Smith, N.; Vidler, E.; Westmarland, L. *Creating Citizen–Consumers: Changing Publics and Changing Public Services*; Sage: London, UK, 2007.
44. Seyfang, G. Shopping for sustainability. *Environ. Politics* **2005**, *14*, 290–306. [CrossRef]
45. Hoogendoorn, G.; Sütterlin, B.; Siegrist, M. The climate change beliefs fallacy: The influence of climate change beliefs on the perceived consequences of climate change. *J. Risk Res.* **2020**. [CrossRef]
46. de Andalucia, J. Andalucía Innova. Especial Cambio Climático. Available online: http://www.andaluciainvestiga.com/revista/pdf/100PreguntasMedioAmbiente/100medioambiente.pdf (accessed on 7 May 2020).
47. European Environment Agency. Final Energy Consumption by Sector. Available online: http://www.eea.europa.eu/data-and-maps/indicators/final-energy-consumption-by-sector-5/assessment#toc-1 (accessed on 8 May 2020).
48. Whitmarsh, L. Behavioural responses to climate change: Asymmetry of intentions and impacts. *J. Environ. Psychol.* **2009**, *29*, 13–23. [CrossRef]
49. Circular Economy: Definition, Importance and Benefits. Available online: https://www.europarl.europa.eu/news/en/headlines/economy/20151201STO05603/circular-economy-definition-importance-and-benefits (accessed on 9 April 2020).

50. European Commission. A New Circular Economy Action Plan for a Cleaner and More Competitive Europe. Available online: https://ec.europa.eu/environment/circular-economy (accessed on 10 April 2020).
51. Gifford, R. The dragons of inaction: Psychological barriers that limit climate change mitigation and adaptation. *Am. Psychol.* **2011**, *66*, 290–302. [CrossRef] [PubMed]
52. Oreg, S.; Katz-Gerro, T. Predicting proenvironmental behavior cross-nationally: Values, the theory of planned behavior, and value-belief-norm theory. *Environ. Behav.* **2006**, *38*, 462–483. [CrossRef]
53. Kollmuss, A.; Agyeman, J. Mind the gap: Why do people act environmentally and what are the barriers to pro-environmental behavior? *Environ. Educ. Res.* **2002**, *8*, 239–260. [CrossRef]
54. Franzen, A. Environmental attitudes in international comparison: An analysis of the ISSP surveys 1993 and 2000. *Soc. Sci. Q.* **2003**, *84*, 297–308. [CrossRef]
55. Hunter, L.M.; Hatch, A.; Johnson, A. Cross-national gender variation in environmental behaviors. *Soc. Sci. Q.* **2004**, *85*, 677–694. [CrossRef]
56. Aoyagi-Usui, M.; Vinken, H.; Kuriyabashi, A. Pro-environmental attitudes and behaviors: An international comparison. *Hum. Ecol. Rev.* **2003**, *10*, 23–31.
57. Gelissen, J. Explaining popular support for environmental protection: A multilevel analysis of 50 nations. *Environ. Behav.* **2007**, *39*, 392–415. [CrossRef]
58. Guerin, D.; Crete, J.; Mercier, J. A multilevel analysis of the determinants of recycling behavior in the European countries. *Soc. Sci. Res.* **2001**, *30*, 195–218. [CrossRef]
59. Gifford, R.; Scannell, L.; Kormos, C.; Smolova, L.; Biel, A. Temporal pessimism and spatial optimism in environmental assessments: An 18-nation study. *J. Environ. Psychol.* **2009**, *29*, 1–12. [CrossRef]
60. Kemmelmeier, M.; Król, G.; Kim, Y.H. Values, economics and proenvironmental attitudes in 22 societies. *Cross-Cult. Res.* **2002**, *36*, 256–285. [CrossRef]
61. Sarigöllü, E. A cross-country exploration of environmental attitudes. *Environ. Behav.* **2009**, *41*, 365–386. [CrossRef]
62. Bord, R.J.; Fisher, A.; O'Connor, R.E. Public perceptions of global warming: United States and international perspectives. *Clim. Res.* **1998**, *11*, 75–84. [CrossRef]
63. Brechin, S.R. Comparative public opinion and knowledge on global climatic change and the Kyoto protocol: The, U.S. against the world? *Int. J. Sociol. Soc. Pol.* **2003**, *23*, 106–134. [CrossRef]
64. Dunlap, R.E.; Mertig, A.G. Global concern for the environment: Is affluence a prerequisite? *J. Soc.* **1995**, *51*, 121–137. [CrossRef]
65. Brechin, S.R.; Kempton, W. Global environmentalism: A challenge to the postmaterialism thesis? *Soc. Sci. Q.* **1994**, *75*, 245–269.
66. Inglehart, R. *Modernization and Postmodernization: Cultural, Economic, and Political Change in 43 Societies*; Princeton University Press: Princeton, NJ, USA, 1997.
67. Franzen, A.; Meyer, R. Environmental attitudes in cross-national perspective: A multilevel analysis of the ISSP 1993 and 2000. *Eur. Sociol. Rev.* **2010**, *26*, 219–234. [CrossRef]
68. Diekmann, A.; Franzen, A. The wealth of nations and environmental concern. *Environ. Behav.* **1999**, *31*, 540–549. [CrossRef]
69. Schultz, P.W.; Zelezny, L. Values as predictors of environmental attitudes: Evidence for consistency across 14 countries. *J. Environ. Psychol.* **1999**, *19*, 255–265. [CrossRef]
70. Washington, H.; Cook, J. *Climate Change Denial: Heads in the Sand*; Earthscan: London, UK; New York, NY, USA, 2011.
71. Cook, J.; Nuccitelli, D.; Green, S.A.; Richardson, M.; Winkler, B.; Painting, R.; Way, R.; Jacobs, P.; Skuce, A. Quantifying the consensus on anthropogenic global warming in the scientific literature. *Environ. Res. Lett.* **2013**, *8*, 024024. [CrossRef]
72. Smith, E.K.; Mayer, A. A social trap for the climate? Collective action, trust and climate change risk perception in 35 countries. *Glob. Environ. Chang.* **2018**, *49*, 140–153. [CrossRef]
73. Bain, R.; Cronk, R.; Hossain, R. Global assessment of exposure to faecal contamination through drinking water based on a systematic review. *Trop. Med. Int. Health* **2014**, *19*, 917–927. [CrossRef]
74. Tranter, B.; Booth, K. Scepticism in a changing climate: A cross-national study. *Glob. Environ. Chang.* **2015**, *33*, 154–164. [CrossRef]
75. Running, K. World citizenship and concern for global warming: Building the case for a strong international civil society. *Soc. Forces* **2013**, *92*, 377–399. [CrossRef]

76. Kvalřy, B.; Finseraas, H.; Listhaug, O. The publics' concern for global warming: A cross-national study of 47 countries. *J. Peace Res.* **2012**, *49*, 11–22. [CrossRef]
77. Pew Research Center. Pew Global Attitudes & Trends Database. 2017. Available online: http://www.pewglobal.org/question-search/?qid=2863&cntIDs=&stdIDs (accessed on 11 June 2020).
78. Ding, D.; Maibach, E.; Zhao, X. Support for climate policy and societal action are linked to perceptions about scientific agreement. *Nat. Clim. Chang.* **2011**, *1*, 462–466. [CrossRef]
79. Swim, J.; Clayton, S.; Doherty, T.; Gifford, R.; Howard, G.; Reser, J.; Stern, P.; Weber, E. Psychology and global climate change: Addressing a multi-faceted phenomenon and set of challenges. In *A Report by the American Psychological Association's Task Force on the Interface between Psychology and Global Climate Change*; American Psychological Association: Washington, DC, USA, 2009.
80. Weber, E.U.; Stern, P.C. Public understanding of climate change in the United States. *Amer. Psychol.* **2011**, *66*, 315. [CrossRef]
81. Hornsey, M.J.; Harris, E.A.; Fielding, K.S. Relationships among conspiratorial beliefs, conservatism and climate scepticism across nations. *Nat. Clim. Chang.* **2018**, *8*, 614–620. [CrossRef]
82. Roppolo, M. Americans More Skeptical of Climate Change Than Others in Global Survey. CBS News 23 July 2014. Available online: https://www.cbsnews.com/news/americans-moreskeptical-of-climate-change-than-others-in-global-survey (accessed on 12 June 2020).
83. Hamilton, L.C. *Public Awareness of the Scientific Consensus on Climate*; SAGE: London, UK, 2016; pp. 1–11. [CrossRef]
84. Nisbet, M. The IPCC Report Is a Wake Up Call for Scholars, Advocates, and Philanthropists. Article Adapted from a Lecture Delivered October 11 at the University of Pennsylvania. 2018. Available online: https://medium.com/wealth-of-ideas/the ipcc-report-isa-wake-up-call-for-scholars-advocates-and-philanthropists-36415d4882f (accessed on 14 June 2020).
85. Scruggs, L.; Benegal, S. Declining public concern about climate change: Can we blame the great recession? *Glob. Environ. Chang.* **2012**, *22*, 505–515. [CrossRef]
86. Fairbrother, M.; Sevä, I.J.; Kulin, J. Political trust and the relationship between climate change beliefs and support for fossil fuel taxes: Evidence from a survey of 23 European countries. *Glob. Environ. Chang.* **2019**, *59*, 102003. [CrossRef]
87. Lee, S.; Paik, H.S. Korean household waste management and recycling behavior. *Build. Environ.* **2011**, *46*, 1159–1166. [CrossRef]
88. Schwartz, S.H. Normative influences on altruism. In *Advances in Experimental Social Psychology*; Berkowitz, L., Ed.; Academic Press: New York, NY, USA, 1977.
89. Uniting Church. For the Sake of the Planet and All Its People. Uniting Church of Australia Statement. Available online: http://www.unitingjustice.org.au/environment/uca-statements/item/481-for-the-sakeof-the-planet-and-all-its-people (accessed on 11 June 2020).
90. Catholic Earthcare Australia. Climate Change—Our Responsibility to Sustain God's Earth. PositionPaper from Catholic Earthcare Australia. Available online: http://www.catholicearthcare.org.au/POSITION_PAPER.html (accessed on 21 June 2020).
91. Bremer, J.; Linnenluecke, M.K. Determinants of the perceived importance of organisational adaptation to climate change in the Australian energy industry. *Aust. J. Manag.* **2017**, *42*, 502–521. [CrossRef]
92. Bockarjova, M.; Steg, L. Can protection motivation theory predict pro-environmental behavior? Explaining the adoption of electric vehicles in the Netherlands. *Glob. Environ. Chang.* **2014**, *28*, 276–288. [CrossRef]
93. Bergantino, A.S.; Catalano, M. Individual's psychological traits and urban travel behavior. *Int. J. Transp. Econ.* **2016**, *43*, 341–359.
94. De Young, R. Expanding and evaluating motives for environmentally responsible behavior. *J. Soc.* **2000**, *56*, 509–526.
95. Frick, J.; Kaiser, F.G.; Wilson, M. Environmental knowledge and conservation behavior: Exploring prevalence and structure in a representative sample. *Pers. Individ. Differ.* **2004**, *37*, 1597–1613. [CrossRef]
96. Mobley, C.; Vagias, W.M.; DeWard, S.L. Exploring additional determinants of environmentally responsible behavior: The influence of environmental literature and environmental attitudes. *Environ. Behav.* **2010**, *42*, 420–447. [CrossRef]
97. Hines, J.M.; Hungerford, H.P.; Tomera, A.N. Analysis and synthesis of research on responsible environmental behavior: A meta-analysis. *J. Environ. Educ.* **1986**, *18*, 1–8. [CrossRef]

98. Diamantopoulos, A.; Schlegelmilch, B.B.; Sinkovics, R.R.; Bohlen, G.M. Can socio-demographics still play a role in profiling green consumers? A review of the evidence and an empirical investigation. *J. Bus. Res.* **2003**, *56*, 465–480. [CrossRef]
99. McCright, A.M.; Dunlap, R.E. Cool dudes: The denial of global climate change among conservative white males in the United States. *Glob. Environ. Chang.* **2011**, *21*, 1163–1172. [CrossRef]
100. Whitmarsh, L.; O'Neill, S. Green identity, green living? The role of pro-environmental self-identity in determining consistency across diverse pro-environmental behaviours. *J. Environ. Psychol.* **2010**, *30*, 305–314. [CrossRef]
101. Ajzen, I.; Gilbert, N. Attitudes and the prediction of behavior. In *Attitudes and Attitude Change*; Crano, W.D., Prislin, R., Eds.; Psychology Press: New York, NY, USA, 2008; pp. 289–311.
102. Zelezny, L.C.; Chua, P.; Aldrich, C. Elaborating on gender differences in environmentalism. *J. Soc.* **2000**, *56*, 443–457.
103. KatrienVan, A.; Holvoet, N. Intersections of Gender and Marital Status in Accessing Climate Change Adaptation: Evidence from Rural Tanzania. *World Dev.* **2016**, *79*, 40–50.
104. Smucker, T.A.; Wisner, B.; Mascarenhas, A.; Munishi, P.; Wangui, E.; Sinha, G. Differentiated livelihoods, local institutions, and the adaptation imperative: Assessing climate change adaptation policy in Tanzania. *Geoforum* **2015**, *59*, 39–50. [CrossRef]
105. Mohai, P.; Twight, B.W. Age and environmentalism: An elaboration of the Buttel model using national survey evidence. *Soc. Sci. Q.* **1987**, *68*, 798–815.
106. Kaiser, F.G.; Keller, C. Disclosing situational constraints to ecological behavior: A confirmatory application of the mixed Rasch model. *Eur. J. Psychol. Assess.* **2001**, *17*, 212–221. [CrossRef]
107. Xiao, C.; McCright, A. Environmental concern and socio-demographic variables: A case study of statistical models. *J. Environ. Educ.* **2007**, *38*, 3–13. [CrossRef]
108. Tjernström, E.; Tietenberg, T. Do differences in attitudes explain differences in national climate change policies? *Ecol. Econ.* **2008**, *65*, 315–324. [CrossRef]
109. Butler, J. *Gender Trouble: Feminism and the Subversion of Identity*; Routledge: London, UK, 1990.
110. Dai, J.; Kesternich, M.; Löschel, A.; Ziegler, A. Extreme weather experiences and climate change beliefs in China: An econometric analysis. *Ecol. Econ.* **2015**, *116*, 310–321. [CrossRef]
111. Cottrell, S.P. Influence of sociodemographics and environmental attitudes on general responsible environmental behavior among recreational boaters. *Environ. Behav.* **2003**, *35*, 347–375. [CrossRef]
112. Dietz, T.; Stern, P.C.; Guagnano, G.A. Social structural and social psychological bases of environmental concern. *Environ. Behav.* **1998**, *30*, 450–471. [CrossRef]
113. Zsóka, A.; Szerényi, Z.; Széchy, A.; Kocsis, T. Greening due to environmental education? Environmental knowledge, attitudes, consumer behaviour and everyday proenvironmental activities of Hungarian high school and university students. *J. Clean. Prod.* **2012**, *48*, 126–138. [CrossRef]
114. Francis, B. Teaching manfully? Exploring gendered subjectivities and power via analysis of men teachers' gender performance. *Gend. Educ.* **2008**, *20*, 109–122. [CrossRef]
115. Nayak, A.; Kehily, M.J. Gender undone: Subversion, regulation and embodiment in the work of Judith Butler. *Br. J. Sociol. Educ.* **2006**, *27*, 459–472. [CrossRef]
116. Chant, S. Women-headed households: Poorest of the poor? Perspectives from Mexico, Costa Rica and the Philippines. *IDS Bull.* **1997**, *28*, 26–48. [CrossRef]
117. Englert, B. Changing land rights and gendered discourses: Examples from the Uluguru Mountains. In *Women's Land Rights and Privatization in Eastern Africa*; Tanzania, B., Englert, E., Daley, Eds.; James Currey: Oxford, UK, 2008; pp. 83–100.
118. Boto-García, D.; Bucciol, A. Climate change: Personal responsibility and energy saving. *Ecol. Econ.* **2020**, *169*, 106530. [CrossRef]
119. Subaru, K.M.; Bertone, E. From Thoughts to Actions: The Importance of Climate Change Education in Enhancing Students' Self-Efficacy. *Aust. J. Environ. Educ.* **2019**, *35*, 123–144.
120. Adger, N.W. Social vulnerability to climate change and extremes in coastal Vietnam. *World Dev.* **1999**, *27*, 249–269. [CrossRef]
121. Berman, R.J.; Quinn, C.H.; Paavola, J. Identifying drivers of household coping strategies to multiple climatic hazards in Western Uganda: Implications for adapting to future climate change. *Clim. Dev.* **2015**, *7*, 71–84. [CrossRef]

122. Clark, C.F.; Kotchen, M.J.; Moore, R.M. Internal and external influences on pro-environmental behavior: Participation in a green electricity program. *J. Environ. Psychol.* **2003**, *23*, 237–246. [CrossRef]
123. Dimian, G.; Marin, E.; Jablonsky, J. Investigating the long and short-run salary-employment relationship in Romania: A sectorial approach using the ARDL model. *Econ. Comput. Econ. Cybern. Stud. Res.* **2019**, *53*, 5–20.
124. Bene, C. Are fishers poor or vulnerable? Assessing economic vulnerability in small-scale fishing communities. *J. Dev. Stud.* **2009**, *45*, 911–933. [CrossRef]
125. Domene, E.; Sauri, D. Urbanisation and water consumption: Influencing factors in the metropolitan region of Barcelona. *Urban. Stud.* **2006**, *43*, 1605–1623. [CrossRef]
126. Renwick, M.E.; Archibald, S.O. Demand side management policies for residential water use. *Land Econ.* **1998**, *74*, 343–359. [CrossRef]
127. Gouldson, A.; Colenbrander, S.; Sudmant, A.; Papargyropoulou, E.; Kerr, N.; McAnulla, F.; Hall, S. Cities and climate change mitigation: Economic opportunities and governance challenges in Asia. *Cities* **2016**, *54*, 11–19. [CrossRef]
128. Zhifu, M.; Guan, D.; Liu, Z.; Liu, J.; Viguié, V.; Fromer, N.; Wang, Y. Cities: The core of climate change mitigation. *J. Clean. Prod.* **2019**, *207*, 582–589.
129. Rosenzweig, C.; Solecki, W.; Hammer, S.A.; Mehrotra, S. Cities lead the way in climate-change action. *Nature* **2010**, *467*, 909–911. [CrossRef]
130. Vermeer, M.; Rahmstorf, S. Global sea level linked to global temperature. *Proc. Natl. Acad. Sci. USA* **2009**, *106*, 21527–21532. [CrossRef] [PubMed]
131. Reckien, D.; Salvia, M.; Heidrich, O.; Church, J.M.; Pietrapertosa, F.; De Gregorio-Hurtado, S.; D'Alonzo, V.; Foley, A.; Simoes, V.; Krkoška, V.; et al. How are cities planning to respond to climate change? Assessment of local climate plans from 885 cities in the EU-28. *J. Clean. Prod.* **2018**, *191*, 207–219. [CrossRef]
132. Unsworth, K.L.; Fielding, K.S. It's political: How the salience of one's political identity changes climate change beliefs and policy support. *Glob. Environ. Chang.* **2014**, *27*, 131–137. [CrossRef]
133. Scott, D.; Willits, F.K. Environmental attitudes and behavior: A Pennsylvania survey. *Environ. Behav.* **1994**, *26*, 239–260. [CrossRef]
134. Leiserowitz, A.A.; Maibach, E.; Roser-Renouf, C.; Hmielowski, J.D. Politics & Global Warming: Democrats, Republicans, Independents, and the Tea Party. *Sci. News* **2011**, 137. [CrossRef]
135. Leviston, Z.; Walker, I. Beliefs and denials about climate change: An Australian perspective. *Ecopsychology* **2012**, *4*, 277–285. [CrossRef]
136. Pidgeon, N. Public understanding of, and attitudes to, climate change: UK and international perspectives and policy. *Clim. Policy* **2012**, *12*, S85–S108. [CrossRef]
137. Ziegler, A. Political orientation, environmental values, and climate change beliefs and attitudes: An empirical cross country analysis. *Energy Econ.* **2017**, *63*, 144–153. [CrossRef]
138. Steg, L. Limiting climate change requires research on climate action. *Nat. Clim. Chang.* **2018**, *8*, 754–761. [CrossRef]
139. Harring, N. Understanding the effects of corruption and political trust on willingnessto make economic sacrifices for environmental protection in a cross-national perspective. *Soc. Sci. Q.* **2013**, *94*, 660–671. [CrossRef]
140. Kollmann, A.; Reichl, J. How trust in governments influences the acceptance of environmental taxes. In *Political Economy and Instruments of Environmental Politics*; Schneider, F., Kollman, A., Reichl, J., Eds.; MIT Press: Cambridge, UK, 2015.
141. Klenert, D.; Mattauch, L.; Combet, E.; Edenhofer, O.; Hepburn, C.; Rafaty, R.; Stern, N. Making carbon pricing work for citizens. *Nat. Clim. Chang.* **2018**, *8*, 669–677. [CrossRef]
142. Pharr, S.J.; Putnam, R.D.; Dalton, R. A Quarter-Century of Declining Confidence. *J. Democr.* **2000**, *11*, 5–25. [CrossRef]
143. Dalton, R. The Social Transformation of Trust in Government. *Int. Rev. Sociol.* **2005**, *15*, 133–154. [CrossRef]
144. Lo, A.; Chow, A. The Relationship between Climate Change Concern and National Wealth. *Clim. Chang.* **2015**, *131*, 335–348. [CrossRef]
145. Burton-Chellew, M.N.; May, R.M.; West, S.A. Combined inequality in wealth and risk leads to disaster in the climate change game. *Clim. Chang.* **2013**, *120*, 815–830. [CrossRef]

146. Brown, T.C.; Kroll, S. Avoiding an uncertain catastrophe: Climate change mitigation under risk and wealth heterogeneity. *Clim. Chang.* **2017**, *141*, 155–166. [CrossRef]
147. Lankao, P.R.; Nychka, D.; Tribbia, J.L. Development and greenhouse gas emissions deviate from the 'modernization'theory and 'convergence'hypothesis. *Clim. Res.* **2008**, *38*, 17–29. [CrossRef]
148. Parry, M.; Canziani, O.; Palutikof, J.; van der Linden, P.; Hanson, C. Climate Change 2007: Impacts, Adaptation and Vulnerability. In *Contribution of Working Group II to the Fourth Assessment Report of the Intergovernmental Panel on Climate Change*; Parry, M.L., Canziani, O.F., Palutikof, J.P., van der Linden, P.J., Hanson, C.E., Eds.; Cambridge University Press: Cambridge, UK, 2007; p. 976.
149. Anderson, K. Talks in the city of light generate more heat. *Nature* **2015**, *528*, 437. [CrossRef]
150. Anderson, K.; Le Quéré, C.; Mclachlan, C. Radical emission reductions: The role of demand reductions in accelerating full decarbonization. *Carbon Manag.* **2014**, *5*, 321–323. [CrossRef]
151. Lamb, W.F.; Steinberger, J.K. Human well-being and climate change mitigation. *Wiley Interdiscip. Rev. Clim. Chang.* **2017**, *8*, e485. [CrossRef]
152. Adua, L.; York, R.; Schuelke-Leech, B.A. The human dimensions of climate change: A micro-level assessment of views from the ecological modernization, political economy and human ecology perspectives. *Soc. Sci. Res.* **2016**, *56*, 26–43. [CrossRef]
153. Mol, A.P.J.; Sonnenfeld, D.A.; Spaargaren, G. *The Ecological Modernisation Reader: Environmental Reform in Theory and Practice*; Routledge: London, UK; New York, NY, USA, 2009.
154. Balogun, A.L.; Marks, D.; Sharma, R.; Shekhar, H.; Balmes, C.; Maheng, D.; Salehi, P. Assessing the Potentials of Digitalization as a Tool for Climate Change Adaptation and Sustainable Development in Urban Centres. *Sustain. Cities Soc.* **2020**, *53*, 101888. [CrossRef]
155. United Nations. The Age of Digital Interdependence. Available online: https://www.un.org/en/pdfs/DigitalCooperation-report-for%20web.pdf (accessed on 28 June 2020).
156. Kalas, P.P.; Finlay, A. *Planting the Knowledge Seed: Adapting to Climate Change Using ICTs*; Building Communication Opportunities (BCO) Alliance, 2009; Available online: https://www.apc.org/sites/default/files/BCO_ClimateChange.pdf (accessed on 7 May 2020).
157. Vining, J.; Ebreo, A. What makes a recycler? A comparison of recyclers and nonrecyclers. *Environ. Behav.* **1990**, *22*, 55–73. [CrossRef]
158. Lawrence, E.K.; Estow, S. Responding to misinformation about climate change. *Appl. Environ. Educ. Commun.* **2017**, *16*, 117–128. [CrossRef]
159. Gullberg, A.T.; Aardal, B. Is climate change mitigation compatible with environmental protection? Exploring voter attitudes as expressed through "old" and "new" politics in Norway. *Environ. Policy Gov.* **2019**, *29*, 67–80. [CrossRef]
160. Reckien, D.; Flacke, J.; Dawson, R.J.; Heidrich, O.; Olazabal, M.; Foley, A.; Hamann, J.J.-P.; Orru, H.; Salvia, M.; Hurtado, S.D.G.; et al. Climate change response in Europe: What's the reality? Analysis of adaptation and mitigation plans from 200 urban areas in 11 countries. *Clim. Chang.* **2013**, *122*, 331–340. [CrossRef]
161. Puppim Oliveira, J.A.; Doll, C.N.H.; Kurniawan, T.A.; Geng, Y.; Kapshe, M. HuisinghPromoting win–win situations in climate change mitigation, local environmental quality and development in Asian cities through co-benefits. *J. Clean. Prod.* **2013**, *58*, 1–6. [CrossRef]
162. Bulkeley, H.K. Kern Local government and the governing of climate change in Germany and the UK. *Urban. Stud.* **2006**, *43*, 2237–2259. [CrossRef]
163. Heidrich, O.; Tiwary, A. Environmental appraisal of green production systems: Challenges faced by small companies using life cycle assessment. *Int. J. Prod. Res.* **2013**, *51*, 5884–5896. [CrossRef]
164. Hunt, A.; Watkiss, P. Climate change impacts and adaptation in cities: A review of the literature. *Clim. Chang.* **2011**, *104*, 13–49. [CrossRef]
165. Villarroel, R.; Walker, M.B.; Beck, J.W.; Hall, R.J.; Dawson, O. Identifying key technology and policy strategies for sustainable cities: A case study of London. *Environ. Dev.* **2017**, *21*, 1–18. [CrossRef]
166. Wende, W.; Bond, A.; Bobylev, N.; Stratmann, L. Climate change mitigation and adaptation in strategic environmental assessment. *Environ. Impact Assess. Rev.* **2012**, *32*, 88–93. [CrossRef]
167. Baranzini, A.; Carattini, S. Effectiveness, earmarking and labeling: Testing the acceptability of carbon taxes with survey data. *Environ. Econ. Policy Stud.* **2017**, *19*, 197–227. [CrossRef]

168. Human Development Report 2019 beyond Income, beyond Averages, beyond Today: Inequalities in Human Development in the 21st Century. Available online: http://hdr.undp.org/sites/default/files/hdr2019.pdf (accessed on 12 May 2020).
169. Special Eurobarometer Report 489. Available online: https://op.europa.eu/en/publication-detail/-/publication/85980533-adbc-11e9-9d01-01aa75ed71a1 (accessed on 12 June 2020). [CrossRef]
170. Special Eurobarometer Report 490. Available online: https://ec.europa.eu/clima/sites/clima/files/support/docs/report_2019_en.pdf (accessed on 12 June 2020). [CrossRef]

Article

Circular Economy in Poland: Profitability Analysis for Two Methods of Waste Processing in Small Municipalities

Przemysław Zaleski [1,2] and Yash Chawla [1,*]

1 Department of Operations Research and Business Intelligence, Wroclaw University of Science and Technology, Wyb. Wyspianskiego 27, 50-370 Wroclaw, Poland; przemyslaw.zaleski@pwr.edu.pl

2 Department of Finance and Strategic Analysis, EkoPartner Recykling Ltd., ul. Zielona 3, 59-300 Lubin, Poland

* Correspondence: yash.chawla@pwr.edu.pl; Tel.: +48-693-290-935

Received: 18 July 2020; Accepted: 22 September 2020; Published: 4 October 2020

Abstract: The problem of diminishing resources on our plant is now getting due attention from the governments as well as scientists around the world. The transition from a linear economy to a circular economy (CE) is now among the top priorities. This article discusses the implementation of the circular economy paradigm in Poland through the analysis of the existing and planned mechanisms, and actions taken by the Polish government which can be replicated by other young European countries. Further, the article discusses the direction of change and projected measures planned by the Polish government to improve the quality of municipal solid waste management. In this context, profitability analysis is carried out for two methods of waste processing (incineration and torrefaction) intended for small municipalities and settlements in which district heating and trading of generated electricity are not feasible. The results of the analysis shows that torrefaction is clearly a more desirable waste processing option as a step towards the implementation of CE for civic society in the urban context, as well as profitability, in comparison to incineration. The analysis accounts for several scenarios before the lockdown caused due to the COVID-19 pandemic and after it was lifted.

Keywords: circular economy; profitability analysis; municipal solid waste processing; incineration; waste recycling; torrefaction; COVID-19

1. Introduction

The rapidly rising social awareness appeals for changes in the approach towards the use of resources in the widely understood economy and industry. Concern for the resources on our planet is undoubtedly a serious matter, as their volume continues to shrink while the demand continues to grow, due to a constantly increasing population and modernization of lifestyle. Projections by the United Nations (UN) show that the current global population stands at around 7.7 billion and that it will continue to grow to nearly 10 billion by the year 2050 [1,2]. According to the World Wide Fund For Nature [3], assuming that all people on the planet have the same lifestyle as in the European Union (EU), human beings would consume ecosystem resources intended for the whole year by May 10th [4]. The intergovernmental Science-Policy Platform on Biodiversity and Ecosystem Services (IPBES) document [5] discusses the highly negative impact of economic development on the environment and ecosystems. In particular, the areas occupied by cities have nearly doubled since 1992. Almost 33% of land areas and 75% of potable water resources are used for plant and animal production, while plastic pollution has increased from 4.86 (urban population 39%) billion in 1980 to 7.8 billion (urban population 56%) in 2020 [6]. This situation calls for a shift in the strategic perception and immediate changes in the approach towards a responsible use of natural resources, including secondary resources and recycling of waste. In order for the changes to be effective, they must cover a wide range of entities and processes in various dimensions—including construction, production and technology, material, organizational, institutional, political, economic and socio-cultural aspects. This is emphasized in many scientific publications dedicated to the circular economy (CE) [7–11].

In 2015, all UN member states voted for the adoption of the 2030 Agenda for Sustainable Development, which highlights 17 Sustainable Development Goals (SDGs) and specific targets for each goal [12]. Among these 17 SDGs, Goal 12: ensure sustainable consumption and production patterns and Goal 11: make cities and human settlements inclusive, safe, resilient and sustainable, are directly related with the theme addressed in this article, whereas, Goal 7: ensure access to affordable, reliable, sustainable and modern energy for all and Goal 9: to build resilient infrastructure, promote inclusive and sustainable industrialization and foster innovation, are also somewhat related to it. The idea of a CE is not new, since various solutions have already existed in the scientific and economic literature, e.g., the concept of sustainable development [13,14]. Researchers have argued that, even if well-off countries incur high spending to protect the environment, the relative effects of their actions would be rather weak [15]. The current linear model of the economy is a hindrance towards solving the problem of waste, CO_2 emission or extraction of natural resources, as it contradicts the business interests of the global economic powers. Moving together towards a CE model, which guarantees sustainability and competitiveness simultaneously, would be the most effective solution [9,15]. CE opens vast opportunities for various kinds of businesses and, increasing the material circularity within the economy, can also alleviate poverty, but systemic and disruptive changes would only take effect if significant changes to the existing regulatory structures were carried out [16].

The European Union (EU) adopted a CE Action Plan, in December 2015, to steer the existing economic development model towards a more sustainable one. This action plan mandates a continued commitment of all EU stakeholders for a transition to a more circular economy, where the value of products, materials and resources would be maintained as long as possible, and the generation of waste would be minimized [17]. In 2019, 4 years after its adoption, the action plan was finally considered to be fully complete and consists of 54 actions which have been delivered or are being implemented [18]. The 54 sections have been divided into various categories, such as: production, consumption, waste management, market for secondary raw materials, and sectoral action for plastics, food waste, critical raw materials, bio-mass and bio-based materials, innovation and investments, and monitoring. For waste management, one of the first actions was to revise the legislation enforcing the principles of the CE to the existing waste management policies. Through the directive (EU) 2018/851 of the European Parliament and the Council of 30 May 2018, the previous Directive 2008/98/EC on waste was amended [19]. Since this directive, different measures have been taken by the EU member countries and researchers have carried out various studies in this regard as well. In terms of municipal waste management, it has been found that the Central and Northern EU members are among the best performers, whereas, the worst are Eastern European countries [20]. The barriers found for waste management in Eastern as well as some Central European countries include: a focus on low-cost options, vast discrepancies of waste management performance across different regions and lack of cooperation between different layers of multi-governance in waste management [21]. Hence, it is safe to say that these countries would be more inclined to adopt a waste processing method following the CE principle, if there are higher profits or financial incentives involved as compared to the existing methods. This study, through the profitability analysis on empirical data, would recommend one such waste processing method, a vital practical implication.

As a member state of the European Union, Poland is obligated to implement and transpose all legal solutions established by the EU institutions onto the national legislation. However, not without a critical assessment. Polish postulates submitted to the European Commission in September 2015 addressed four important issues [22]: (i) support for innovative initiatives; (ii) consideration of the service sector in the CE implementation; (iii) enhanced flow of raw materials; and (iv) an improvement of their quality by means of sustainable production and consumption. In January 2016,

Poland announced its position in which it supported shifting to CE, however, under certain conditions enabling adjustment of goals to the capacity of individual member states. The formal state authority, responsible for the implementation of actions and preparation of Poland to the transformation of the Polish economy to CE is the Ministry of Development which, pursuant to the national legislation, appointed the Task Force for CE composed of the representatives of ministries involved in the economic transformation. The effect of works of the Task Force was publishing—as early as December 2016—of the draft concerning the Polish Circular Economy Roadmap. The direction of actions presented in the document is de facto compliant with the understanding of CE described above. As stated in the roadmap: "The concept of sustainable production is based, not only on the principle of increasing the resource productivity that is decreasing the volume of raw materials consumed for a unit of produced goods, but also on anticipating the reduction of negative environmental impact of the production processes, including, in particular, the context of reducing greenhouse gas emissions and volume of produced waste" [23]. This is why we focus in this article on waste management, in particular, municipal waste management.

There has been prior research on municipal waste management, and sustainable model solutions for it have also been proposed in Poland, other parts of EU and the world [20,24,25]. There has been intensive progress in Poland for waste management, however, studies in the literature do not address the cost effectiveness option, which is one of the barriers towards implementation of CE in this sector [21]. There is a lack of empirical studies which discuss the profitability of adopting CE concepts for waste processing, especially in this part of Europe. One of the reasons for this is the limited access to actual market data used by the waste management companies, which is usually guarded as a business secret. With this article, we wish to address this gap in the literature. In general, the aim of our research is to elaborate on the implementation of CE in Poland, one of the younger EU countries but also one of the largest member states, in terms of municipal waste management and to draw out the examples which could be useful for other young EU countries, especially the Visegrad four (Czech Republic, Hungary, Poland and Slovakia). Among the Visegrad four, Poland is leading the way in waste management [26,27], especially in terms of policies, and still there is scope for improvement. In particular we aim to carry out the profitability analysis, on empirical data, for two methods of waste processing, incineration and torrefaction, intended for small municipalities and settlements in Poland in which district heating and trading of generated electricity are not feasible, to show which method would be more beneficial in terms of profitability as well as CE concept.

The structure of the article is as follows. After giving a brief introduction to the article in Section 1, findings from the literature regarding municipal waste management, implementation of circular economy and actions taken by Polish government towards implementation of CE, are elaborated in Section 2. This is followed by the background of the case study and the description of data and methods used for the profitability analysis of two methods of municipal solid waste management, incineration and torrefaction, in Section 3 which include the results of the case study in Section 2. The case study accounts for the conditions that existed prior to the lock-down caused due to the COVID-19 pandemic and after the re-opening, in Poland. This is followed by the conclusions in Section 4 and lastly, Section 5 highlights the limitations of the current study and the future research horizons it opens up.

2. Circular Economy and Municipal Waste Management in Poland

2.1. Circular Economy—Actions Taken by Polish Government

Scientists, businesses and governments around the world have discussed various roadmaps towards achieving circular economy in general as well as waste management in particular [10,28–31]. Poland also drafted its own CE roadmap after a critical analysis of the prevaling conditions and the new directive by the European Commission [23]. This draft of the Polish CE Roadmap was included as one of the elements of the Strategy for Responsible Development (the official strategy of the Polish government) and submitted for inter-ministerial and social consultations. The document consists of four parts concerning the identified economic issues (or actions): sustainable industrial production, sustainable consumption (SC), bioeconomy and new business models. There is a large interest among young EU countries, to integrate the idea of the economy in a closed system, to waste management, as a positive correlation between waste generation and recycling rates has been observed [32]. The main aim here is to minimize the amount of waste generated and to manage the resulting waste in accordance with waste hierarchy (waste prevention, preparation for re-use, recycling, other recovery, disposal). According to the implementation of the idea of the economy in a closed system, there is an obligation to collect detailed data on waste management. In Poland, the Central Statistical Office is responsible for keeping these detailed records. It prepares reports based on the "Waste Catalog", dividing the waste into groups, subgroups and types, depending on the source of their generation [33]. The Sections 2.1.1–2.1.4 elaborate the four important actions towards CE by the Polish government and some related actions or literature from other countries, especially the Visegrad four.

2.1.1. Promotion of Life Cycle Assessment (LCA)

Implementation of control measures at each stage of the value chain and environmental impacts are not novelties in science, however, the CE perspective provided this approach with a more comprehensive method of production settlement. For example, see [34], where the authors compared the productivity in different production systems: conventional and innovative, with the use of various indicators. They presented direct, indirect and total emissions in a life cycle, recovered waste, consumption of primary resources, as well as carbon dioxide emission maps, presenting the effects for the entire supply chain. From the literature, it is evident that, an inclusion of CE principles to a sustainable supply chain management may give noticeable benefits for both, environmental protections and market volatility perspectives [34,35]. According to CE principles, product information disclosure must contain data concerning the chemical composition, decomposition and environmental impact throughout the entire life cycle. This applies both, to the cycle, understood as use period, and to all or selected elements related to acquisition of resources, transport, production process, using and withdrawal from the market [36]. Thus, the assessment covers the calculation of the volume of raw materials and energy consumers, as well as emissions into the environment. In a critical global review of LCA in municipal waste management Europe and Asia to be leading in such studies, whereas 178 countries having no published studies in this regard since 2013 [37]. In Poland LCA is a part of Environmental Protection Law, State Ecological Policies, Strategies for implementations of integrated product policy and changing production and consumption models leading to more sustainable and circular economy. LCA is also an integral part of waste management, assessment of new technologies, public procurements and eco-labeling [38,39]. Moreover LCA is also an evaluation criterion for access to public subsidies from structural funds, such as the Innovative Economy Operations Program. One of the cases for application of LCA in Poland was in the analysis of the life cycle of industrial water meters [40]. Complete information on the production of water meters, which covered factors including, e.g., supply of raw materials, transportation, production, its use and so on, were documented on balance tables for a set of quantities and, as a result, components were created. Based on these assessments, the criteria for the acceptance of industrial water meters were agreed upon. Another example is the study by [41], which compared the LCA of current and future electricity generation systems in Poland and Czech Republic. Various LCA models have been used to assess the municipal waste management in Poland which have lead to positive recommendations towards higher efficiency of waste management [39,42,43]. As compared to Poland, such assessments for municipal waste management have not been carried out in young EU countries, especially the Visegrad four [37].

2.1.2. Sustainable Consumption (SC)

Over the past several years, the European member states have formed various strategies to promote SC. Most of these strategies concentrate on improvements in technology for production and products itself and have little or no focus on the consumers' consumption practices. Nevertheless, communication with the consumers, through labels or information campaigns, to steer them towards more sustainable behaviour has been carried out to certain extent [44]. In Poland, the essence of the second action on SC, was to implement any and all activities aiming at changing the lifestyle of citizens including, primarily, the approach to widely understood consumption of tangible goods. Both the relative absence of wars, and large economic disasters in recent years, had contributed to a dynamic growth of consumption throughout Europe. Due to the COVID-19 pandemic, the subsequent lock-down and the on going after effects has brought about a change in the consumers' consumption patters [45–47]. Scientists have raised the question whether COVID-19 would actually aid in increasing SC or due to the after effects, such as social distancing, would actually cause a negative effect on SC [48]. This remains to be seen in the coming years. Prior to the COVID-19 pandemic, the scientists studying SC reported that following the other EU Member States, Poland recorded a several percent growth in GDP in the period of 2003–2012 and, thus, a growth in consumption expenditures related to the general increase in wealth of the society [49]. However, the education level did not keep pace with consumption levels and the threats posed by this process to the natural environment. Change in the approach to consumption is neither swift nor easy, however, it can be supported by technological changes that take place globally throughout the world. For instance, Genus [50] describes the technological changes, for example: common access to the Internet, digitisation; and sociological changes, such as direct communication between the clients and fashion designers or change of attitude to "do it yourself" (DIY), which reduces mass production in favour of one's own production.

Issues, emphasized in the scientific literature, translate into political actions. This is why the communication of the European Commission clearly states that the choices made by the consumers may support CE development or impede it [17]. SC is supposed to satisfy basic human needs and, at the same time, minimize consumption of natural resources and reduce production of waste and emissions. The actions taken by Poland in this scope address the CE concept and the process itself accelerates by regulations in three core areas [22]:

- Municipal waste management by mandatory segregation into fractions (paper, plastics, glass, municipal waste)
- Disposal of so-called bulk waste to the city collection points
- Collection of used batteries in dedicated containers

Other regulations for the implementation of selective collection of hazardous waste from households, not covered by the existing collection system, are under planning. Another crucial element is the economic incentive applied by the government for the citizens to move towards CE. For instance, in the form of the My Current programme [51], the households receive reimbursement of partial costs for installation of solar cells, which promotes renewable electricity generation. The government had also lowered the excise tax for electric vehicles to 0%, encouraging the citizens to transition from conventional fuel vehicles to electric ones [52]. The government also initiated the Clean Air Program, which offers co-financing for replacing old and inefficient heat sources with solid fuel for modern heat sources that meet the highest standards, as well as carrying out accompanying thermo modernization works of the building [53]. The Clean Air program budget is set at the level of 103 billion PLN, with an implementation period until 2029. An equally important initiative, being undertaken for addressing the CE, is raising awareness and actions on food wastage. According to research performed for the Federation of Polish Food Banks in 2018, Poles waste nearly 235 kg of food per annum [54]. The change effort concentrates primarily on raising awareness on the consumer side, as building internal inhibitions for the consumer and society is the most effective long term solution.

2.1.3. Bioeconomy

EU has been leading the way when it comes to transitioning to bioeconomy, especially through the initiatives under the Horizon 2020 programme [55]. Poland on its road to implement CE, has taken several steps towards a bioeconomy, i.e., biological cycles in the economy [56]. To better understand the idea behind this term, one should differentiate between two cycles:

- The first applies to the essence of biological or natural raw material—in most cases—from which the product is made and which is a renewable resource
- The second applies to its technical aspect, including construction, functionality and technological quality, which is a non-renewable resource

Pursuant to the CE concept, both conditions must be present and have the so-called circularity value. In most cases, we can speak about bioeconomy in the context of renewable resource management options, where the resource is reusable and recoverable in the form of biomass. The agenda for bioecnomy in the Polish CE roadmap includes two aspects. The first about general actions aiming towards creating suitable conditions for the implementation and development of bioeconomy in Poland and the second about actions for specific areas such as creating local value chains for industrial sector in general and for the power sector in particular [23]. Application of bioeconomy in specific activities is reasoned in the industrial sectors with organic consumption, such as food, feed, wood and timber, cellulose and paper, pharmaceuticals, textiles, furniture, fisheries, agricultural and biofuel industries [57]. To date, biomass in Poland has been used primarily in the furniture sector as a raw material for the production of boards or energy raw material for the co-incineration process as an effective component to reduce CO_2 emissions. In terms of municpal waste management, in bioeconomy, the organic waste is considered to be a raw material which can further utilized to obtain useful output, such as value added-biofuels/chemicals from waste [58]. Both the methods of waste processing, incineration and torrefaction, analysed in this article are suitable under the bioeconomy concept.

Under the various actions taken in Poland towards development of bioeconomy, it has been recommended to develop a database assessing the potential of biomass of plant, as well as animal origin and biodegradables for the Polish economy as a whole. Assessing the economic and environmental impact in the individual sectors would be of key importance. Establishing the system for identification, reporting, statistics on individual types of biomass, along with coding and an assessment of production capacity, is also planned. This database will underpin the platform for, alongside cooperation with the industry, an establishment of bioeconomic clusters and coordinators at the inter-sectoral and supraregional level, aiding in eliminating the barriers for optimum economic development. Experts analyzing the role of the bioeconomy show its ability to expand in the total economy [59]. Reports have shown that Poland is among the top countries in Europe when it comes to bioeconomy. Among the Visegrad group, Poland is leading the way with bio based fuel, bioenergy, biomass processing and conversion, other bio industries such as biorefineary, biochemicals and biopharmaceuticals, whereas the others have made some progress in agro-food sector [60]. Nevertheless, there is still vast room for improvement for Poland for achieving a circular bioeconomy.

2.1.4. New Business Models

The fourth action, taken for implementation of the CE, was to specify the conditions for establishment of new business models. The design of the latter was based on combining the supply of value to the customer with simultaneous consideration of the so-called closing the loop concept, for example: designing of the rules for 'circular' logistics that will consider the entire flow of raw materials, semi-products and products from the producer by transport to the final consumer. While establishing the new business models, focus was on close cooperation between the producers from the same sector (which has been a standard for years) as well as in different sectors, such as cooperation at the level of raw material, technology, transport (car) sharing, sharing of production tools, premises, regeneration of used parts, human resources and comprehensive economic symbiosis [61]. It has also been stressed in the CE roadmap of Poland that involvement of social cooperatives, associations, foundations and personnel would be utmost important for the new business models. The organizations and people have an indepth understanding of the local communities due to which they could simultaneously contribute to economic, environmental as well as social objectives. To adapt to the new business models, incentives have also been offered as a motivation. The primary incentive included tax deductions, simplified accounting procedures, easy use of waste as a secondary raw material with various economic incentives and simplification of administrative decisions. Although actions have and are being taken, the available scientific literature emphasizes that, to accelerate the implementation of new business models, it is necessary to constantly identify them and study promising new tools and processes [62]. Case studies, such as the one in this artcile, give validation to adoption of new and advanced methods for businesses, which would be a step towards CE as well as prove to be more profitable.

Another important element of the road map is the extended producer responsibility (EPR), which is an important component of sustainable production [63–69]. It obliges the producer to collect and manage waste produced from the products sold by them in the market. In general, EPR results in the implementation of the polluter pays principle [70], which makes the producers consider the whole life cycle of the raw materials at the production planning stage. It also forces the producers to use solutions that would enable collection of waste and the highest possible recycling rate. In the EU legislation, the extended producer responsibility was implemented under Article 8 of the Waste Framework Directive (2008/98/EC) and incorporated by the Directive of the European Parliament and of the Council (EU) of 30 May 2018 [71]. In Poland, the EPR principle has been currently applied to packaging, end-of-life vehicles, electrical and electronic equipment, tyres, batteries and lubricating oils. While preparing the roadmap, the task force recommended to extend the currently valid provisions by imposing obligations on the producers. It was highlighted that the structure of provisions should be in a way that would encourage them to take care of the natural environment and reduce the volume of waste production, but, in such a manner as not to impede on business activity. The applicable legal solutions were planned to be implemented in a short time perspective, since Poland and the other EU Member States were obliged to deliver the goals planned at the level of the EU legislation. The planned objectives also set targets of recycling for packaging, of 65% of recycling by 2025 and 70% by 2030. The Task Force document also recommended comprehensive review of the legal acts in force, to check the extent to which they cover the product life cycle, with the EPR definition acting as a reference point and the Minister of Environment being the authority responsible for review. The review was to be supplemented with analysis of risks and benefits to the entrepreneurs. It should be emphasized here that, with regard to the objectives, Poland immediately expressed its positive attitude by considering them in the Resolution of the Council of Ministers no. 88 of 1 July 2016 [72].

Summarizing the actions taken by Polish government towards implementation of CE, detailed in Sections 2.1.1–2.1.4, the following Table 1 is presented. These steps could be replicate by the other young EU countries, especially the Visegrad four, in their efforts towards transition towards CE.

Although a number of steps are being taken in the director of implementing CE concept in Poland, there is still a wide scope for improvement of structure implementations, especially in the municipal waste management sector [25].

Table 1. Summary of important actions taken in Poland towards implementation of CE [23].

1	Reviewing and updating of regulations in production and packaging sectors, to adapt them to the EU requirements and orient them towards implementation of CE.
2	Carrying out S.W.O.T. (Strength, Weakness, Opportunities, Threats) analysis for control and reporting of extended producer responsibility, followed by development of solutions to eliminate irregularities.
3	Running information campaign on benefit to the businesses' social image by fulfilling extended produced responsibility.
4	Adapting the methodologies developed by European Commission to develop information package for calculating the environmental impact of products and economic activities.
5	Developing recommendations to amend national regulations of municipal waste management, to adapt to CE concepts, through constant study of the effectiveness and efficiency of the existing regulations.
6	Recording all the municipal waste flows, especially the ones which have been unrecorded so far, in order to boost the recovery and recycling in the waste management sector.
7	Running social campaigns to promote sustainable consumption practices.
8	Developing government information platform on CE for citizens and businesses.
9	Building a formal and permanent team at governmental agencies responsible for individual areas of bioeconomy which would take up define the further direction for bioeconomy and supervise its implementation.
10	Review and analysis of regulations and supply potential at local and national level for biomass, preceded by building and adopting suitable methodology.
11	Carrying out comprehensive analysis of Research, Development and Innovation priorities for implementation of bioeconomy in Poland.
12	Scouting the local value chains and exploring the feasibility of establishment of bio-refineries.
13	Studying the feasibility of updating the tax system, in order to provide boost competitiveness of businesses abiding by the concept of CE.
14	Proposing legislation which would promote sharing of property, real estate and movable, especially for short term leasing of vacant passenger transport and residential spaces.
15	Proposing and updating the law of public procurement.
16	Creating a support ecosystem for businesses to enable them to transit to CE model.
17	Creating an internet platform encompassing multiple industry sectors for easy lending and sharing of products with low usage frequency.
18	Promoting introduction of CE in research programmes and curriculum at the university through an incentive system.
19	Monitoring through the "oto-GOZ" ["this-is-CE"] project (the Gospostrateg programme).

2.2. *Municipal Waste Generation and Management in Poland*

In 2016, Poland's industrial production amounted to 23.4%, compared to the average EU share of 17.4% [73] and Poland is being looked up to as Europe's new growth engine [74]. In 2004-2016 the average growth rate of the industrial production in Poland was 5.3% compared to 0.5% in the EU [73]. In terms of waste production, as per the data provided by Statistics Poland [75], the main sources of waste in Poland are listed below and also elaborated in Figure 1:

- Mining and extraction (nearly 48% of total produced waste)
- Industrial processing (23.8%)
- Energy production and supply (14.35%).
- Municipal waste (9.78%)
- Others (4.07%)

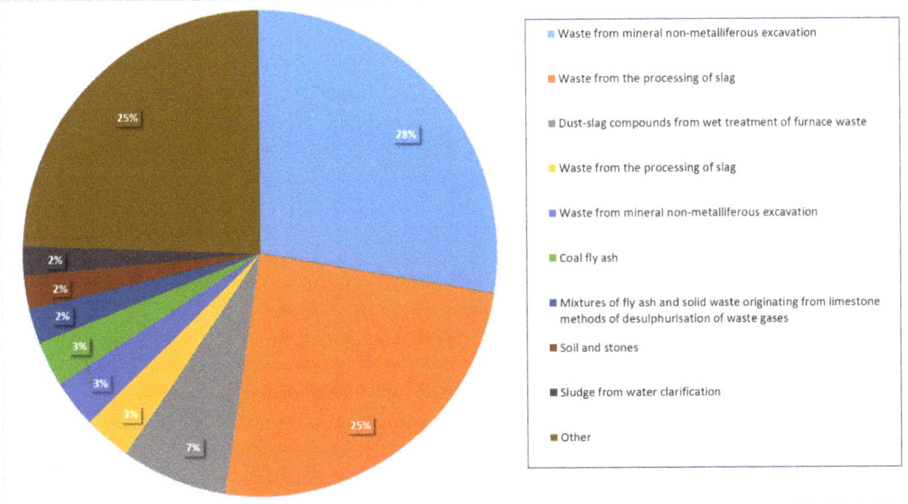

Figure 1. Structure of waste generated in Poland in 2017, by waste type and excluding municipal waste. Data source: Statistics Poland [75].

In terms of waste management, Poland faces challenges, for both industrial waste and a growing volume of municipal waste. Pursuant of the Polish Waste Act [76], municipal waste is interpreted as: "waste produced in households, excluding end-of-life vehicles, as well as waste containing no hazardous waste from the other waste producers, which, due to its nature or composition, is similar to waste produced in households; non-segregated (mixed) municipal waste remain as non-segregated (mixed) municipal waste even if processed, which did not change its properties significantly". According to Statistics Poland data, in 2017, Poland generated 12 million tonnes of municipal waste, of which 6.8 million tonnes (57%) was recycled, 5 million (42%) for land-filling (EU target by 2035 is ≤10% [77]), while 1% (nearly 0.2 million tonnes) was combusted without energy recovery [75]. The recycling activities included :

- recycling at the level of 3.2 million tonnes (27%)
- thermal processing with energy recovery at the level of 2.7 million tonnes (22%)
- composting at the level of 848 thousand tonnes (7%) [78]

As of 2018, Poland was the second lowest producer of municipal waste in EU (329 kg per capita), which is considerably lower than the average of EU28 (489 kg per captia) [77]. Figure 2, shows the stand of other EU member states in terms of kilogram of municipal waste generated per capita in 2005 and 2018. It can be seen that Poland has been successful in reducing its municipal waste by over 50% as compared to 2005. In the EU, only Romania recorded lower municipal waste volume as compared to Poland in 2018. There has been a steep increase in the municipal waste volume due to the COVID-19 pandemic, the affects of which has strained the waste management systems and has presented a massive challenge for the municipalities [79,80].

Directive (EU) 2018/851 of the European Parliament and of the Council of 30 May 2018 amending Directive 2008/98/EC on waste, set forth common recycling targets, for EU member states. Recycling 85% paper, 55% plastic, 60% aluminium, 80% ferrous metal and 75% glass is expected to be achieved by 2030, as per the directive. This required a major overhaul of the waste collection systems (WCSs) as the firs step, because of the need to segregate waste and divert materials to appropriate destinations for recycling [81]. In response to this, major investments have been made by the EU member states towards updating the existing WCS to meet the requirements. Poland also took the necessary steps in this direction.

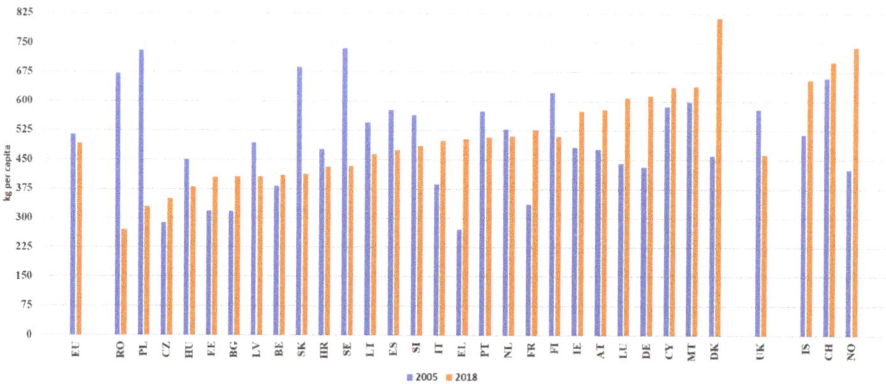

Figure 2. Municipal waste generation in the European Union (EU) 1995 and 2018 [77].

Polish municipal waste management system has seen a drastic change since the 1990s, when the responsibility of waste disposal shifted form the municipal authorities to the house owners, under the Poland's privatization program. This was governed by the Act of 13 September 1996 on maintaining cleanliness and order in communes, and was amended several times. It also completely changed the system of fees for waste management. It became the responsibility of the house owners to organize waste collection, obligatory inclusion of all municipal residents in the municipal waste collection system and imposing on them the obligation to segregate all types of waste: biodegradable (green), paper, glass, plastics and municipal waste, and then deliver them into the regional municipal waste treatment installations (RIPOKs—Regionalnej Instalacji Przetwarzania Odpadów Komunalnych) (art. 9e of the Act). Subsequently four other acts were passed: (i) the Act on maintaining cleanliness and order in municipalities (2012); (ii) the Act on waste (2012); (iii) the Act on management of packaging and packaging waste (2013); and (iv) Act of 23 January 2020 amending the act on waste and certain other acts. The responsibility of collection and disposal of municipal was was again transferred to the municipalities and also brought made it obligatory for waste handlers to act in consistence with the national standards. Moreover to encourage the recovery and recycling, the fees on landfills was also hiked by imposing additional taxes [82]. This was in line with the EU's landfill directive which required all member to introduce higher taxed on landfill so as to encourage recycling [83]. A clear co-relation between higher taxes and decrease in the percentage of waste being landfill was observed and reported for the EU-28 [83], which made this a popular choice of strategy among the member states. Among the Visegrad four, apart from Poland, Czech Republic, Hungary and Slovakia have not yet made strides for implementation such strategies in their municipal waste management policies [84].

The present municipal waste management system, in Poland, is established on the basis of the National Waste Management Plan. It specifies the operation (obligations and tasks) of the entities acting as the RIPOKs. Majorly, these are self-governmental or private installations (currently approx. 200). Information on the operation of these entities are submitted to the Central Waste System, which handles all information related to waste management. To enable assessment of their operation, waste management is performed for a strictly defined area of activity (region), where the installation can collect and transport waste only within the voivodeship on which it performs the activity. It should be noted that transport of municipal (mixed) and green waste, outside the region, is currently illegal [76].

To date, the system anticipated that the RIPOKs have a sufficient capacity to accept and process waste from the area inhabited by at least 120 thousand inhabitants, meeting the requirements of the best available technique or technology [85]. Difficulties resulting from the inability to manage a significant volume of municipal waste pose a challenge related to their management under the CE paradigm. This applies to non-recyclable calorific fractions, which are fed in a form of alternative fuel to cement factories and residual waste fed to the incineration plants. These include waste fraction of calorific value above > 6 MJ/kg, which, pursuant to the Polish legislation, are to be disposed through land-filling. In the process of adopting the shift towards CE, one of the preferable solutions was the construction of advanced waste incineration plants that have treated a vast majority of waste, along with industrial installations capable of treatment (incineration) in high temperatures or cement factories. At present, there are seven municipal waste thermal processing installations (incineration plants) operating, while construction of the other two are at the final stage: in Rzeszów to be launched in 2019 and in Gdańsk to be launched in 2021 [86,87]. Table 2, shows more details about these nine waste incineration installations in Poland.

Table 2. Municipal waste incineration installations in Poland, as of October 2019.

Site	**Productivity [Thousand mg/Annum]**	**Power Production Capacity [MWe]**	**Thermal Energy Production Capacity [MWt]**
Poznań	210	15	34
Kraków	220	8	35
Bydgosz	180	9.2	27.7
Szczecin	150	13	34
Konin	94	6.75	15.4
Białystok	120	8.68	17.5
Warsaw [1]	330	20	60
Gdańsk [2]	160	30	–
Rzeszów [2]	100	15	7.84

[1] Undergoing modernization; [2] Under construction.

Nominal capacity of the existing municipal waste incineration plants is approx. 1.3 Mt/a, i.e., megaton per annum, while incineration of fractions from municipal waste residue accounts for nearly 0.25 Mt/a and co-incineration of waste fuels, in the cement factory, is ca. 1.6 Mt/a. Considering the volume of municipal waste increasing on a year-to-year basis and including the unspecified consumption by economic migrants from Ukraine (approx. 1.2 million people [88] multiplied by the waste production per capita), demand for non-recyclable fraction treatment poses a serious problem. Poland presents a relatively restrictive approach to the emission of pollution in the process of energy carriers' incineration, regardless of whether these are fossil fuels or municipal waste of non-recyclable fraction coded 19-12-12 [89]. The Polish Ordinance on the emission standards for certain types of installation, fuel incineration sources and waste incineration or co-incineration installations follows the EU Directive 2010/75/EU on industrial emissions to the Polish legislation. Notably, the Directive describes the rules of control and monitoring of pollution emission, exploitation conditions, reporting, conditions and applications for the entities applying for the emission rights. Emission standards, compliant with the Ordinance, are presented in Table 3.

Table 3. Emission standards for installations for the thermal treatment of municipal waste.

Pollution Components	Average Daily Limits [mg/m^3_u]
Dust	10
Hydrogen chloride	10
Hydrogen fluoride	1
Sulphur dioxide	50
Nitrogen oxides *	200
Nitrogen oxides **	400
Carbon oxide	50
Total organic carbon	10
Heavy metals and their compounds ***	0.05

Data source: Ordinance of the Polish Minister of Environment of 1 March 2018. * For installations incinerating more than 6 mg per hour. ** For installations incinerating less than 6 mg per hour. *** Cadmium and Thallium, Mercury.

According to the data in Table 3, while maintaining the emission standards, the use of waste treatment as one of the carriers for energy and heat production, is beneficial to the Polish economy. It is estimated that incineration of 1 mg (Megagram) of municipal waste produces approx. 400 kWh of electric energy and 6.6 GJ of thermal energy. The principles of the use of waste is noticeable in the official communication of the government, indicating that the CE waste is a potential resource and should be used as a material. In this context, waste treatment (waste landfilling, incineration without energy recovery) is treated as a loss of resources and manifestation of economic ineffectiveness by resource wasting. Therefore, resource incineration, without energy production, cannot be treated as compliant with CE. However, the rule of thermal processing of waste with energy recovery only applies as a complement to the municipal waste management system, contributing to reduce their volume on landfills. The levels of reuse and recycling remain unaffected by this. About 30% of the total municipal waste undergoes thermal processing.

The National Fund for Environmental Protection and Water Management, a subordinate authority to the Ministry of Environment, initiated a consultation programme, in the form of sectoral conferences, was initiated. This resulted in a decision to use waste heat where there is infrastructure, that enabled all-year consumption of the produced thermal energy, and addressed the CE concept to form a kind of 'completion' in the municipal waste management system. The governmental agency published an announcement that it observed a growing demand of installations for handling a calorific fraction of waste at the thermal processing installations dedicated to municipal waste (with co-generation of thermal energy) [90]. The agency offers financial support to the initiatives aiming at a modernisation of local heat sources in the context of incineration of waste, under the Sustainable Waste Management programme [91], which is a positive and important step forward towards CE.

The CE roadmap [23] also further outlines the direction of actions and responsibilities of central administration authorities for their implementation. First of all, performing the capacity analysis for this area, along with preparation of the proposals for legislative amendments to increase the economical use of incineration by-products (IBPs) was considered a priority. Factors, such as determination of quality requirements (including environmental requirements) to be met by IBPs and creating the conditions for their use (including in the scope of ecodesign) were also considered in this. Such an approach is compliant with CE and may increase availability of raw materials for other economic sectors, while, at the same time, decreasing the volume of waste managed by landfilling. In addition, IBPs can be successfully used in the construction and road sector as a component for production of concrete blocks, access roads, embankments or other construction layers. The authority designated as the leading authority in the preparation of regulations for the roadmap is the Minister responsible for energy in cooperation with the Minister of Environment and the Minister responsible for construction, spatial planning and development and housing. Supervision over the operational implementation rests upon the Inspection for Environmental Protection [23].

3. Case Study: Profitability Analysis for Two Methods of Municipal Waste Processing

3.1. Background of the Case Study

As of 2019, 38.4 million people live in Poland, producing an average of 330 kg municipal waste per year, which is 12.67 million kg of municipal waste per year in the whole country. There also should be added to these calculations 1.2 million immigrants from Ukraine who, due to their incomplete stay, generate less waste, as indicated in Table 4. The exact numbers, due to the increase in volume of waste generated due to the COVID-19 pandemic, remains to be seen.

Table 4. Volume of mixed municipal waste in Poland per annum. (Data source: Statistics Poland).

Category	Value
Number of inhabitants	38.4 million
Waste volume per capita	0.33 mg/person/a
Annual waste volume	12.67 million mg/a
Number of immigrants	1.2 million
Waste volume per capita (immigrants)	0.2 mg/person/a
Waste produced by migrants	0.24 million mg/a
Total waste volume in 2019	12.91 million mg/a

The thesis that, by far, the greatest impact on the amount of waste generated is influenced by economic factors, especially wealth of the residents, is confirmed by studies which were already carried out in 1996 and continued in the following years [92–95]. According to the experts, it can be assumed that GDP growth by 3.5 percent, will result in a 1.5% increase in municipal waste [96]. Based on the forecasts of the National Bank of Poland, projections of decrease in Poland's population according to the projection by UN as well as Statistics Poland, and together with immigrants from Ukraine [88,97] an extrapolation of the volume of municipal waste in Poland, has been prepared and shown in Table 5. Even though the number of inhabitants are projected to decrease the waste generated per capita is projected to increase. Hence, increasing the total amount of waste generated. The recent COVID-19 pandemic has also affected the municipality waste sector adversely in terms of recycling. It is not that there has just been an increase in the municipal waste volume, but there has also been a reduction in recycling during the lock-down due to COVID-19 [98].

Table 5. Trend extrapolation of the volume of municipal waste in Poland.

Year	Population (in Millions)	Increase in Waste Volume [mg/Person/a]	Annual Waste Volume [Million mg/a]
2019	38.48	0.330	12.7
2020	38.14	0.341	13.0
2021	38.09	0.365	13.9
2022	38.04	0.389	14.8
2023	37.97	0.429	16.3
2024	37.85	0.459	17.4

Considering the declining population of Poland, an the upward trend in the number of economic immigrants from Ukraine and the other countries, is clearly visible [97]. Combining this with the increase in municipal waste due to the higher amount of time spent at home due to the pandemic and its after effect, will make the issue of municipal waste, that cannot be land-filled nor recycled, a more severe problem. In the process of implementing the CE in Poland, the actions aiming at increasing the awareness in the field of waste separation into fractions are required to be taken. In 2018, considering the upward trend in immigration, as well as municipal waste generation, representatives of Regional Municipal Waste Processing Installations (RMWPIs), dealing with waste processing, wrote a letter sent to the Minister of the Environment and pointed the requirement of actions to increase the number of professional recyclers and to use other treatment technologies, e.g., torrefaction [99].

In an interesting publication of scientists from many EU countries dealing with the problem of waste, in the context of the idea of a closed-loop economy, emphasized that the transformation of waste into energy can be one of the key elements of a CE that allows manufacturers to maintain the value of products, materials and resources on the market for as long as possible, minimizing waste and resources [21]. As we show in Table 5, municipal waste has significant volume to be managed and also has the potential to contribute towards CE, as well as generate economic benefits through the use of other waste management methods. The interest of doing so certainly exists in Poland. The current focus of the Polish municipal waste management sector is towards increasing the capacity of municipal waste processing through incineration plants [86,87,100]. Incineration is one step further towards CE but still lacks the ability to fully address the CE concept. During incineration of municipal waste, energy is recovered but there is residue in form of sewage sludge (SSA). Ordinance of the Polish Minister of the Environment of May 11, 2015 on the recovery of waste outside installations and equipment, allows for the recovery of the mineral fraction from the SSA. Recently, research was conducted into the possibilities of using ash from SSA produced in incineration plants as a secondary source of phosphorus (P), which resulted from European Union (EU) legislation that indicated that phosphorus is a critical raw material (CRM) [101]. This residue can be used to prepare eco-friendly cement [102]. However, these methods require additional processes and setup for recovering usable material from the residue. This issue can be over by implying torrefaction process instead of incineration. Studies have shown the torrefaction of mixed municipal waster would yield energy and usable fertilizer or fuel [103,104]. Torrefaction is also referred to as roasting i.e., the process of thermal and chemical processing of organic compounds in specific thermal conditions. In most cases, the literature provides the following parameters: temperature of 200–300 C, heating rate on the inside of the reactor <50 °C/min, input dwell time in the reactor 15–60 min, no oxygen, atmospheric pressure, the effect of which is the production of biocarbon [105,106]. Additionally, this technology is more economically advantageous for companies dealing with waste management in smaller towns.

By 31/01/2020, the Ministry of Climate (because it took over the "waste portfolio" as part of the division of competences between the Ministry of the Environment and the newly created Ministry of Climate in the Government of Poland) had started to update (on a national scale) the list of enterprises appearing in the Provincial Waste Management Plans (WPGO—Wojewódzkich Planów Gospodarki Odpadami). Thus, new companies, which have already started to work on building new incinerators, will appear in the list of WPGO, but the possibilities of applying for new incinerators will not be available. It results from the power of art. 35 b of the Waste Act, added by the Act of 19 July 2019, amending the Act on maintaining cleanliness and order in municipalities and amending certain other acts (Journal of Laws of 2019, item 1579). It reads: thermal treatment of waste will be allowed only in installations specified in the regulation issued by the Minister of Climate. Pursuant to Article 35 b para. 3 of the aforementioned Act, if the installation intended for the thermal transformation of municipal waste or waste from the treatment of municipal waste was not included in the list, i.e., new building permits, integrated permits or permits for processing waste in this installation, shall be refused. Considering that the investment process itself from the preparation of technical documentation to construction lasts on average of 3–4 years, it can be assumed that these installations will not be built anytime soon. Moreover, due to the COVID-19 pandemic further delays have taken place. The pandemic influenced the decision of the government administration, namely the Ministry of the Environment, responsible for issuing administrative decisions regarding the construction of incineration plants in Poland. In accordance with the Polish law, incineration plants can be built only after entering them into the National Waste Management Plan, and then placing them in the Provincial Waste Management Plan (Article 186 of the Act—Environmental Protection Law; i.e., Journal of Laws of 2018, item 799), that clearly states that an investment that is not in accordance with WPGO cannot be implemented. Construction of new incineration plants was suspended due to COVID-19, and the local municipalities from Kraków, Tarnów, Zamość and Żywiec, Rybnik, and Wrocław logeed compaliants agains the decision and are making efforts to overturn it. So far, the Ministry of the Environment has not supported the construction of new installations, emphasizing that this is contrary to EU policy. The construction of new installations would increase the possibility of collecting waste by another 500 to 700 thousand tons, depending on the financial capabilities of investors.

Through empirical evidences from the Polish municipal waste management market, this study compares the profitability of two methods of municipal waste management, incineration, and torrefaction. The time of implementation from the approval to the first operation, cost of implementation, cost of waste processing, revenue from waste processing as well as revenue from the sale of byproduct has been taken into account for the analysis. The conclusions drawn from this analysis would add to the gap found in the literature regarding the empirical analysis of these two methods of waste processing. They would also serve as a recommendation for the policy makers and business investors in the field of municipal waste management regarding the choice between the implementation of incineration or torrefaction from the CE and profitability perspective

3.2. Material and Methods

In the Polish waste collection market, the local governments are responsible in accordance with the law of waste management (Act of 13 September 1996 on maintaining cleanliness and order in municipalities), to organize tenders for waste collection. The RIPOKs, which have the possibility to manage and sort individual waste fractions, place their bid on these tenders. The price from the auction is transferred to contracts concluded for one or two years of waste collection from the Commune. These companies analyze the possibilities of utilizing the oversize fraction through mechanical-biological processes or combustion in existing installations of this type. These companies, RIPOKs, do not have complicated methodologies supported by complex mathematical formulas or models, rather they rely more on their managers' updated market knowledge and their experts. Forecasting of the estimated costs and revenue are also not based on advanced models, such as machine learning based approach or multivariate time series, due to the small number of data points (one for each year and 10 in total, without a uniform time interval). Hence the estimates used by these companies, for contesting the tenders, rely more on the market awareness of the managers. The empirical data used for analysis in this study were collected by one of the authors, as the Director of Sales Department at EkoPartner Lubin (one of the RMWPIs in Poland). Following is the description of the the collected data and the steps followed in making the calculations used for the profitability analysis in this study. Such data and steps are followed by most of the RMWPIs in Poland to make their estimates for the tenders they bid for, hence gives an indepth empirical overview in-terms of profitability analysis:

1. **Data:** Information on the amount of municipal waste generated in Poland is collected daily by an employee of the company from Public Information Bulletins obligatory published on the website of municipal offices, from tender proceedings notices, or from auctions conducted through tender platforms (login trade, allegro, olx, market planet). The data are entered into the databases broken down into semi-annual, annual and two-year contracts.
2. **Data Analysis:** Based on the collected information, employees conduct price analyzes in relation to:

   - the size of the commune

- the amount of waste
- distances to specific locations of municipal waste plants
- the possibility of collection by individual locations—applies to both formal and legal integrated permit defined by volume in tonnes/month as well as real possibilities of mechanical and biological processing
- transport costs
- fixed and variable costs of own (internal) plants

3. **Based on the expert assessment of these data**, scenarios and variants of price forecasts for offer prices from each commune are generated separately.
4. **Assessment of competition:** Price variants are collided with data on competitors, namely:
    - prices offered in tenders for a specific municipality and similar in size
    - the possibilities that competing companies have in terms of collection, storage and storage
    - having own transport
    - human potential (number of brigades/shift)
    - economic and financial potential (a single plant or enterprise belonging to Remondis or Tonsmeier networks).
5. **Final offer evaluation:** Employees prepare final variants of the offer on the basis of their own options and the potential of competition together with the assessment of contract profitability for each commune separately, and then the Management Board of the company decides on the final price or price negotiable at auctions.

Based on the empirical data obtained directly from EkoPartner Lublin, six scenarios (A to F) were built. These scenarios show the forecasts for the amounts of municipal waste in the years 2020–2025 preceded by actual data from 2007 onwards. Initially only three scenarios (A, B and C) based on the amount of waste were forecasted, but due to the impact of COVID-19, revenues from the collection of the oversize fraction increased. There were several overlapping reasons for this increase. The first concerned the presence of a significant number of people at home by switching to home office or being in quarantine, and because of closing schools, colleges and some enterprises (hairdressers, cosmetics, cinemas, theaters, gyms, swimming pools, etc.). Increased consumption turned out to be a natural state, and this proportionally translated into waste production. According to the data from four municipal plants (located in Lower Silesia, Upper Silesia and in central Poland), the average amount of waste collected by these companies, especially municipal waste, i.e., the oversize fraction, increased by 5000. tons per month, i.e., by over 30%. Secondly, the collection price for municipal waste increased by 20%, reaching an average ceiling of PLN 1,000/tonne. This is the effect of both the increase in the amount of waste, but also the problems of many existing municipal waste collection facilities not only limited by limiting the capacity and mechanical and biological treatment of waste, but also from staffing problems—corona viruses and the inability to return some employees from Ukraine, who in some enterprises accounted for up to 60% of employees. Hence, additional three scenarios (D, E and F) which took into account the effects of the pandemic were also taken into consideration.

For Scenario A, the forecasts are based on: the possibility of utilization with the use of existing, as well as, under-construction incineration plants and cement plants; price paths determined by the author based on bilateral public tenders, in which the author participated and the contracts that were concluded by the company. The existing installations, along with the investments in progress—by 2025—will enable for treatment of 14.4 million tonnes of waste, which creates a market gap of nearly 4 million tonnes. Prices of waste, from the oversize fraction up to 2020, come from contracts concluded by the company, and, beyond 2020, are the company's projection. The market value is calculated as: (Annual waste volume—Total Capacity) x Price, while 'Total Capacity' is the sum of Incineration plant capacities, Cement factories, Fraction 0-80 and Raw materials. It is to be noted here that this analysis does not take into account the cost for collection of the municipal waste. These costs are covered in the fees for garbage collection from the residents, hence are not taken into account in the profitability analysis for the torrefying installation. The same principle applies to the receipt of biochar, which is the result of a process at the installation. It is companies interested in biochar that collect raw material from RIPOK with their own transport. Hence for RIPOK both of these costs do not apply. In scenario B, we assume that more incineration plants will be built from 2023, and their utilization capacity will be 2 million tonnes every year, and, in Scenario C, we assume that more incineration plants will arrive, and the utilization capacity will be 2.5 million tonnes from 2023. Scenarios D, E and F are in the same conditions as A, B and C respectively after the effect of real market conditions which arose due to the COVID-19 pandemic.

3.3. *Results of the Case Study*

Table 6 shows the calculations for various waste management options for scenario A, and the market value of the resulting overload fraction of the waste. The amount of waste (overload fraction) that is to be managed has been decreasing steadily since 2007. This is primarily because of the government's initiative to reduce the production of municipal waste. It can be seen, based on the projections, that in the years to come this overload fraction would start to increase again if additional capacity is not added.

Table 6. Waste management options and market value.

Year	Annual Waste Volume [1]	Incineration Plant Capacities [1]	Cement Factories [1]	Fraction 0–80 [1]	Raw Materials [1]	Total [1]	Price [2]	Amount of Waste to be Managed [1]	Market Value [3]
2007	10.67	0.06	1.4	0.0	1.1	1.82	90	8.7	787
2010	10.04	0.06	1.4	0.0	1.6	2.37	100	7.7	768
2012	9.58	0.06	1.4	3.2	1.6	5.58	110	4.0	440
2013	9.47	0.06	1.4	3.2	1.7	5.62	140	3.9	539
2014	10.33	0.06	1.4	5.2	2.0	7.89	140	2.4	342
2015	10.86	0.34	1.4	5.4	2.2	8.64	150	2.2	333
2016	11.68	1.2	1.4	5.8	2.3	10.08	170	1.6	273
2017	11.97	1.2	1.4	6.0	2.4	10.28	180	1.7	304
2018	12.50	1.2	1.4	6.3	2.5	10.65	250	1.9	463
2019	12.7	1.2	1.4	6.3	2.5	10.3	450	1.9	856
2020	13	1.2	1.4	6.5	2.6	10.5	500	2.0	998
2021	13.9	1.2	1.4	6.9	2.8	11.1	525	2.1	1079
2022	14.8	1.6	1.5	7.4	3	11.9	525	2.1	1101
2023	16.3	1.6	1.5	8.1	3.3	12.9	550	2.5	1391
2024	17.4	1.6	1.6	8.7	3.5	13.8	550	2,8	1552
2025	18.3	1.8	1.6	9.1	3.7	14.4	575	3.0	1715

[1] In [million mg/a], [2] In [PLN/mg], [3] In [million PLN].

Considering the five other scenarios, B to F, Figure 3, shows the graph with market values of the overload fraction, in all six scenarios A to F. Projections on the effects of the pandemic clearly show that the overload fraction of the municipal waste is bound to increase and its market value would be even higher. This simply indicates a need for swift action for increasing the processing capacity.

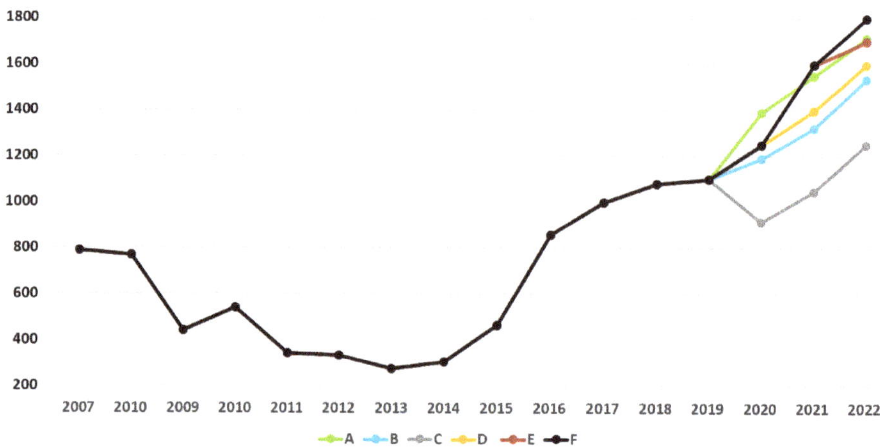

Figure 3. Market values (in million PLN) of the overload fraction to be developed.

Torrefaction and incineration plants have a lot of difference when it comes to implementation time, cost of establishment and waste processing, by products produced and maintenance. Table 7, shows the comparison between various elements for setting up incineration and torrefaction plants for a typical small region in Poland—a pivot with 30,000 tonnes of waste generated on average.

Table 7. Comparison of incineration vs. torrefaction installations.

Characteristic	Incineration	Torrefaction
Implementation timeline	5 years	3 years
Implementation cost	120 million PLN	22 million PLN
Waste processing cost	PLN 350/mg	PLN 110/mg
Generated product	Heat	Fuel
Requirements	Continuous monitoring	No requirements

Data source: EkoPartner.

It can be seen that torrefaction plants are clearly more advantageous as compared to incineration plants. Setting up a torrefaction plant is more than 5 times cheaper as compared to incineration plants, takes 40% less time for implementation, produces the byproduct of fuel and has no additional requirements like continuous monitoring. Moreover, The processing of waste through torrefaction is over 3 times cheaper than incineration, which results in higher profits. Figure 4, shows the comparison between the revenues, costs and profits for processing waste incineration and torrefaction, pre and post the COVID-19 pandemic for processing 30,000 Megagram per annum municipal waste.

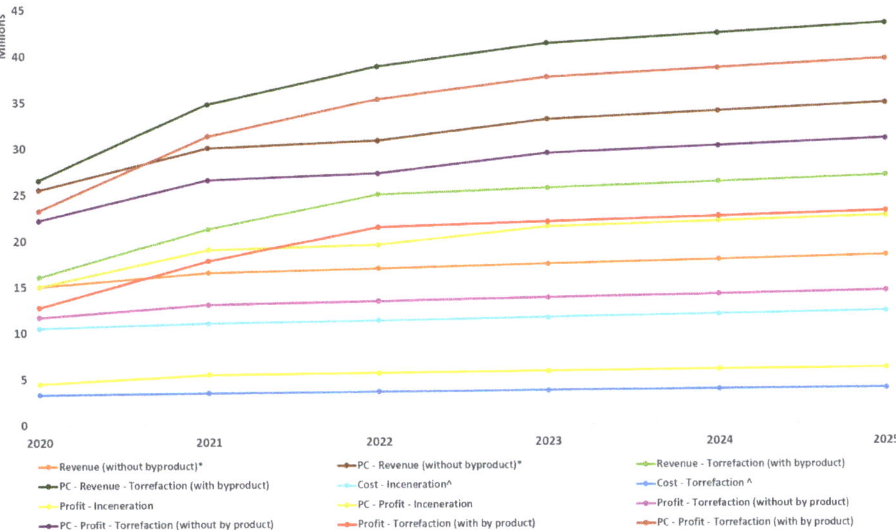

Figure 4. Comparison of revenue, cost and profit for incineration and torrefaction processes of waste processing in Polish municipalities from 2020 to 2025, for 30,000 mg/a municipal waste (data source: EkoPartner). * The revenue from waste collection is the same for both torrefaction and incineration, ^ Cost of processing the was remained same before and after the pandemic, (PC)—Post COVID-19 pandemic.

As shown in Figure 4, the revenues for both, torrefaction and incineration, are the same (shown by the "Revenue (without byproduct)*" line in the graph). The torrefaction process yields biochar as a byproduct, which is a fuel and can be sold directly for additional revenue without any further processing. When the additional revenue from selling the biochar is not considered, the corresponding profits for both the processes are shown by the lines "Profit-Incineration" and "Profit-Torrefaction (without byproduct)" in the graph. Even in this condition (without considering the revenues from byproduct sales), the profit for torrefaction process is higher than that for incineration. If the revenues from the sale of biochar is considered, the revenues and, consequently, the profits, rise (as shown by the lines "Revenue-Torrefaction (with byproduct)" and "Profit-Torrefaction (with byproduct)" respectively). In fact, the projected profits from 2021 onwards for the torrefaction process were even higher than the projected revenues when the sale of byproduct is not considered. This means that only the sale of biochar itself is more than enough for covering the costs of processing the waste. Due to the COVID-19 pandemic the revenue for waste processing increased dramatically, hence increasing the profits of these installations, the effects of which are also shown in the Figure 4. These results show that torrefaction is far more profitable as compared to incineration.

4. Conclusions

The growing environmental challenges and diminishing resources has led countries around the world to take action in moving towards a CE. The concept gains new followers in the world of science, as well as in business and amongst policy-makers. National and regional legislation have been established to facilitate the movement from a linear economy to a CE. COVID-19 pandemic has caused concerns for economies around the world and put severe strain on the resources as the world went under lockdown. Poland, one of the youngest EU countries, had been steadily putting efforts for transitioning towards CE. The Polish roadmap for a CE has introduced new and important elements, such as EPR, which would increase the recycling rates and also waste management. Primarily, the Polish government has taken action towards promoting life cycle assessment, SC, bioeconomy and new business models. It remains to be seen, how these policies are taken forward in the post pandemic era, nevertheless, among the Visegrad four, Poland's actions (summarized in Table 1) can be considered as an example for forming strategies for moving towards CE.

Waste management is one of the challenges face by Poland, both for the industrial waste and the municipal waste. In terms of municipal waste, Poland was able to curtail the volume of municipal waste generate by over 50% as of 2018 in comparison to 1995 and was able to achieve 57% recycling rate in 2017. On the other hand there is still over 42% waste which is being land-filled and 1% combusted without energy recovery, which is against the principle of CE.As per the current capacity of the incineration plants with energy recovery, a little less than 50% of the recycled waste can be processed with thermal recovery and there is a large amount of overload waste fraction. Even before considering the effect of the COVID-19 pandemic, empirical projections showed that this overload waste fraction will grow further, because current incineration plants which are functioning of the ones which would be setup by 2025 shall not be able to cope with the growing municipal waste volume. More so as the people stayed home during the lock-down, caused by COVID-19 and the prevailing work from home policy of a number of organization post the lockdown. RIPOKs in Poland are estimating even higher municipal waste volume from 2021 onwards, which will add further to the overload waste fraction.

As of October 2019, there were six fully functioning municipality incineration installations, one of which is undergoing modernization and two are under construction. Installations of new incineration plants in Poland have to undergo a very length process. The recent amendments and changes at the ministerial level in the Polish government, had complicated things further. Furthermore, the effect of COVID-19 pandemic has delayed the implementation of the new plants which were approved as well as halted the approval process of new projects for such installations. It remains to be seen when the processing of new installations opens up again. Based on the empirical data taken from the actual implemented contracts and the ongoing contract (provided by EkoPartner, one of the RMWPIs in Poland), by 2025, the current incinerators were expected to have a capacity of processing about 14.4 million tonnes of waste, which would have created a market gap of about 4 million tonnes. Due to he prevailing condition this would be a difficult target to meet now, as the implementation period of setting up an incinerator is estimated at 5 years. Hence, any new incinerators (even if approved in 2020 despite the current hold off) could only begin functioning by 2025 and the delay in implementation of already approved projects would create a larger market gap. Torrefaction plants have a shorter implementation period of 3 years, which would prove to be an effective solution in this scenario. If, approved by 2021, these plants can be in service by 2024, and reduce the stress of the overload waste fraction.

The results of the comparison of the profitability analysis of torrefaction and incineration for municipal waste treatment, in this study, further warrants choice of torrefaction. Torrefaction has lower costs of implementation (120 million PLN for incineration as compared to 22 million for torrefaction), which means that five torrefaction plants can be implemented in the cost of 110 million PLN, 10 million PLN less than the cost of one incineration plant. Additionally, they would be ready 2 years earlier. Once implemented, torrefaction would also be more profitable in terms of waste processing, as shown in Figure 4, because of its low operating cost (one third of the cost of incineration). This would also be a strong step towards CE, as the byproduct (biochar—a fuel) does not require any further processing and can be re-introduced in the economy. Considering the revenue obtained from sale of biochar, the profits soar even higher. In fact, the revenue generated from the sale of byproduct would be more than enough to cover the cost for processing the waste from the 2nd year, onwards. This would lead to lower cost of waste processing for municipalities and the saved costs could be diverted towards implementation of more incineration plants. For Poland, it is even more profitable because of the high amount of coal being used in the production of electricity. Biochar is a suitable fuel for the production of electricity, hence it would not just add to the revenues, but also help in reducing emissions, reducing the amount of coal being used and thus leading to decrease in mining. Benefiting from all aspects that required to be added in CE. This fills the gap found in the literature for concert empirical evidence, to concluded that torrefaction is more suitable and profitable option in a CE, as a waste processing option in for processing municipal waste as compared to incineration.

Currently, Poland has not implemented any torrefaction plants on a large scale basis for processing municipal waste. In the current circumstance, post COVID-19 pandemic, governments around the world are striving to strengthen the economies again, using new technologies, systems and solutions. This would also be a conducive time and opportunity for Poland as well as other young EU countries, to consider implementation torrefaction as an effective solution to solve their municipal waste management challenge.

5. Limitations and Future Scope of Research

This study has limitations that point to future works and research avenues. The profitability analysis carried out in this study was based on empirical data from EkoPartner, one of the RMWPIs in Poland. It would be interesting to carry out a study with data from all RWMPIs in Poland. Similar studies can also be carried out in other countries and compared with the current study. It would also be interesting to study the profitability of the torrefaction process, where it has already been implemented, and compare it with the analysis in this study. The preliminary effect of COVID-19 has been accounted for in this study, but it would be interesting to see, in the coming years, on how the CE is implemented and municipality waste management system evolves, in Poland as well as around the world

Author Contributions: P.Z. conceived and designed the research; Y.C. reviewed the research design; P.Z. and Y.C. reviewed the literature; P.Z. gathered the data for the case study, P.Z. and Y.C. carried out the case analysis; P.Z. and Y.C. drafted and edited the paper; P.Z. and Y.C. reviewed the paper. All authors have read and agreed to the published version of the manuscript.

Funding: This work was supported by the Ministry of Science and Higher Education (MNiSW, Poland) core funding for statutory R&D activities.

Acknowledgments: The authors are grateful to Prof. Rafał Weron for comments on an earlier version of the manuscript and to EkoPartner company for providing the data. The authors would also like to acknowledge the reviewers for their constructive remarks, which has greatly help in improving the manuscript.

Conflicts of Interest: The authors declare no conflict of interest.

Abbreviations

The following abbreviations are used in this manuscript:

UN	United Nations
EU	European Union
IBPES	Intergovernmental Science-Policy Platform on Biodiversity and Ecosystem Services
CE	Circular Economy
SDGs	Sustinable Development Goals
CBP	Incineration By-products
EPR	Extended Producer Responsibility
RMWPIs	Regional Municipal Waste Processing Installations

References

1. Lafferty, W.M.; Meadowcroft, J. *Implementing Sustainable Development: Strategies and Initiatives in High Consumption Societies*; Oxford University Press: Oxford, UK, 2000.
2. United Nations. World Population Prospects 2019: Highlights, 2019. Available online: https://www.un.org/development/desa/publications/world-population-prospects-2019-highlights.html (accessed on 26 June 2020).
3. Halkos, G. The Evolution of Environmental Thinking in Economics. Munich Personal RePEc Archive. 2011. Available online: https://mpra.ub.uni-muenchen.de/35580/ (accessed on 18 September 2020).
4. Karamarko, D. A New Kind of Relationship: Supplier and Public Procurer Response to a Circular Economy Insights from Philips Lighting's Deal with Washington Metropolitan Area Transit Authority. Master's Thesis, Faculty of Geosciences, Utrecht University, Utrecht, The Netherlands, 2015.

5. Prieto-Sandoval, V.; Jaca, C.; Ormazabal, M. Towards a consensus on the circular economy. *J. Clean. Prod.* **2018**, *179*, 605–615. [CrossRef]
6. World Population by Year. Available online: https://www.worldometers.info/world-population/world-population-by-year/ (accessed on 17 July 2020).
7. Ying, J.; Li-jun, Z. Study on green supply chain management based on circular economy. *Phys. Procedia* **2012**, *25*, 1682–1688. [CrossRef]
8. Park, J.Y.; Chertow, M.R. Establishing and testing the "reuse potential" indicator for managing wastes as resources. *J. Environ. Manag.* **2014**, *137*, 45–53. [CrossRef]
9. Bonviu, F. The European economy: From a linear to a circular economy. *Rom. J. Eur. Aff.* **2014**, *14*, 78.
10. Merli, R.; Preziosi, M.; Acampora, A. How do scholars approach the circular economy? A systematic literature review. *J. Clean. Prod.* **2018**, *178*, 703–722. [CrossRef]
11. Ngan, S.L.; How, B.S.; Teng, S.Y.; Promentilla, M.A.B.; Yatim, P.; Er, A.C.; Lam, H.L. Prioritization of sustainability indicators for promoting the circular economy: The case of developing countries. *Renew. Sustain. Energy Rev.* **2019**, *111*, 314–331. [CrossRef]
12. United Nations. Sustainable Development Goals. 2015. Available online: https://www.un.org/sustainabledevelopment/ (accessed on 18 September 2020).
13. Jaworski, T.; Grochowska, S. Gospodarka Obiegu Zamkniętego–kryteria osiągnięcia i perspektywa wdrożenia w Polsce. *Arch. Gospod. Odpadami Ochr. Środowiska* **2017**, *19*, 4.
14. Lewandowski, M. Designing the business models for circular economy—Towards the conceptual framework. *Sustainability* **2016**, *8*, 43. [CrossRef]
15. Raftowicz-Filipkiewicz, M. From sustainable development to circular economy. *Econ. Environ. Stud.* **2016**, *16*, 103–113.
16. Berg, A.; Antikainen, R.; Hartikainen, E.; Kauppi, S.; Kautto, P.; Lazarevic, D.; Piesik, S.; Saikku, L. *Circular Economy for Sustainable Development*; *Reports of the Finnish Environment Institute 26*; Finnish Environment Institute: Helsinki, Finland, 2018. Available online: https://helda.helsinki.fi/bitstream/handle/10138/251516/SYKEre_26_2018.pdf (accessed on 18 September 2020).
17. European Commission. Communication from the Commission to the European Parliament, the Council, the European Economic and Social Committee and the Committee of the Regions: Closing the Loop—An EU Action Plan for the Circular Economy. 2015. Available online: https://eur-lex.europa.eu/resource.html?uri=cellar:8a8ef5e8-99a0-11e5-b3b7-01aa75ed71a1.0012.02/DOC_1&format=PDF (accessed on 26 June 2020).
18. European Commission. Commission Staff Working Document, Accompanying the Document: Report from the Commission to the European Parliament, The Council, the European Economic and Social Committee and the Committee of the Regions on the Implementation of the Circular Economy Action Plan. 2019. Available online: https://eur-lex.europa.eu/legal-content/EN/TXT/HTML/?uri=CELEX:52019SC0090&from=EN (accessed on 26 June 2020).
19. Legislation L150. *Off. J. Eur. Union* **2018**, *61*, 109–140.
20. Castillo-Giménez, J.; Montañés, A.; Picazo-Tadeo, A.J. Performance and convergence in municipal waste treatment in the European Union. *Waste Manag.* **2019**, *85*, 222–231. [CrossRef] [PubMed]
21. Malinauskaite, J.; Jouhara, H.; Czajczyńska, D.; Stanchev, P.; Katsou, E.; Rostkowski, P.; Thorne, R.J.; Colon, J.; Ponsá, S.; Al-Mansour, F.; et al. Municipal solid waste management and waste-to-energy in the context of a circular economy and energy recycling in Europe. *Energy* **2017**, *141*, 2013–2044. [CrossRef]
22. Poland'S Comments on the Circular Economy (Non-Paper). 2015. Available online: https://archiwum.mpit.gov.pl/media/26446/CE_nonpaper_Poland.pdf (accessed on 25 June 2020).
23. Roadmap Transformation Project towards a Circular Economy Sent to KRM (In Polish: Projekt Mapy Drogowej Transformacji w Kierunku Gospodarki O obiegu Zamkniętym Przesłany na KRM). Available online: https://archiwum_mpit.bip.gov.pl/projekty-zarzadzen-i-uchwal/projekt-mapy-drogowej-transformacji-w-kierunku-gospodarki-o-obiegu-zamknietym-przeslany-na-krm.html (accessed on 25 June 2020).
24. Perrot, J.F.; Subiantoro, A. Municipal waste management strategy review and waste-to-energy potentials in New Zealand. *Sustainability* **2018**, *10*, 3114. [CrossRef]
25. Smol, M.; Duda, J.; Czaplicka-Kotas, A.; Szołdrowska, D. Transformation towards Circular Economy (CE) in Municipal Waste Management System: Model Solutions for Poland. *Sustainability* **2020**, *12*, 4561. [CrossRef]
26. Waste in the Visegrad Four: Poland Leading the Way. Available online: https://www.euractiv.com/section/circular-economy/news/waste-in-the-visegrad-four-poland-leading-the-way/ (accessed on 27 August 2020).

27. Recycling Rate of Municipal Waste. Available online: https://ec.europa.eu/eurostat/tgm/table.do;jsessionid=9NfjbF_FygohKRFGrq5-0hLvHt9D1iP-0AL8QdXMdpIxKGX0LABW!-755574875?tab=table&plugin=1&language=en&pcode=cei_wm011 (accessed on 27 August 2020).
28. Kalmykova, Y.; Sadagopan, M.; Rosado, L. Circular economy–From review of theories and practices to development of implementation tools. *Resour. Conserv. Recycl.* **2018**, *135*, 190–201. [CrossRef]
29. de Sousa Jabbour, A.B.L.; Jabbour, C.J.C.; Godinho Filho, M.; Roubaud, D. Industry 4.0 and the circular economy: A proposed research agenda and original roadmap for sustainable operations. *Ann. Oper. Res.* **2018**, *270*, 273–286. [CrossRef]
30. Sarc, R.; Curtis, A.; Kandlbauer, L.; Khodier, K.; Lorber, K.; Pomberger, R. Digitalisation and intelligent robotics in value chain of circular economy oriented waste management—A review. *Waste Manag.* **2019**, *95*, 476–492. [CrossRef]
31. Borrello, M.; Pascucci, S.; Cembalo, L. Three propositions to unify circular economy research: A review. *Sustainability* **2020**, *12*, 4069. [CrossRef]
32. Taušová, M.; Mihaliková, E.; Čulková, K.; Stehlíková, B.; Tauš, P.; Kudelas, D.; Štrba, L. Recycling of Communal Waste: Current State and Future Potential for Sustainable Development in the EU. *Sustainability* **2019**, *11*, 2904. [CrossRef]
33. Kulczycka, J.; Dziobek, E.; Szmiłyk, A. Challenges in the management of data on extractive waste—The Polish case. *Miner. Econ.* **2019**, 1–7. [CrossRef]
34. Genovese, A.; Acquaye, A.A.; Figueroa, A.; Koh, S.L. Sustainable supply chain management and the transition towards a circular economy: Evidence and some applications. *Omega* **2017**, *66*, 344–357. [CrossRef]
35. Johnsen, T.E.; Howard, M.; Miemczyk, J. *Purchasing and Supply Chain Management: A Sustainability Perspective*; Routledge: London, UK, 2018.
36. Jaworski, T.J.; Grochowska, S. Circular Economy-the criteria for achieving and the prospect of implementation in Poland. *Arch. Waste Manag. Environ. Prot.* **2017**, *19*, 14.
37. Khandelwal, H.; Dhar, H.; Thalla, A.K.; Kumar, S. Application of life cycle assessment in municipal solid waste management: A worldwide critical review. *J. Clean. Prod.* **2019**, *209*, 630–654. [CrossRef]
38. Kulczycka, J. Life cycle thinking in Polish official documents and research. *Int. J. Life Cycle Assess.* **2009**, *14*, 375–378. [CrossRef]
39. Kulczycka, J.; Lelek, L.; Lewandowska, A.; Zarębska, J. Life Cycle Assessment of Municipal Solid Waste Management - Comparison of Results Using Different LCA Models. *Pol. J. Environ. Stud.* **2015**, *24*, 125–140. [CrossRef]
40. *Eco-efficiency of Technology (In Polish: Ekoefektywność Technologii)*; Kleibera, M., Ed.; Scientific Publisher of the Institute for Sustainable Technologies—National Research Institute (In Polish: Wydawnictwo Naukowe Instytutu Technologii Eksploatacji—Państwowego Instytutu Badawczego): Radom, Poland, 2011.
41. Burchart-Korol, D.; Pustejovska, P.; Blaut, A.; Jursova, S.; Korol, J. Comparative life cycle assessment of current and future electricity generation systems in the Czech Republic and Poland. *Int. J. Life Cycle Assess.* **2018**, *23*, 2165–2177. [CrossRef]
42. Grzesik, K.; Usarz, M. A life cycle assessment of the municipal waste management system in Tarnów. *Geomat. Environ. Eng.* **2016**, *10*, 29–38. [CrossRef]
43. Żygadło, M.; Dębicka, M. The environmental assessment of mechanical-biological waste treatment by LCA method. *Struct. Environ.* **2016**, *8*, 260–265.
44. Wolff, F.; Schönherr, N.; Heyen, D.A. Effects and success factors of sustainable consumption policy instruments: A comparative assessment across Europe. *J. Environ. Policy Plan.* **2017**, *19*, 457–472. [CrossRef]
45. Hall, M.C.; Prayag, G.; Fieger, P.; Dyason, D. Beyond panic buying: Consumption displacement and COVID-19. *J. Serv. Manag.* **2020**. [CrossRef]
46. Chen, H.; Qian, W.; Wen, Q. The Impact of the COVID-19 Pandemic on Consumption: Learning from High Frequency Transaction Data. 2020. Available online: https://ssrn.com/abstract=3568574 (accessed on 18 September 2020).
47. Chen, C.F.; de Rubens, G.Z.; Xu, X.; Li, J. Coronavirus comes home? Energy use, home energy management, and the social-psychological factors of COVID-19. *Energy Res. Soc. Sci.* **2020**, *68*, 101688. [CrossRef] [PubMed]
48. Cohen, M.J. Does the COVID-19 outbreak mark the onset of a sustainable consumption transition? *Sustain. Sci. Pract. Policy* **2020**, *16*, 1–3. [CrossRef]

49. Śleszyńska Świderska, A. Consumption and economic growth in Poland. In the shadow of the global economic crisis. *Nier ó Wno Ś Soc. Econ. Growth* **2014**, *38*, 352–364.
50. Genus, A. *Sustainable Consumption: Design, Innovation and Practice*; Springer: Berlin/Heidelberg, Germany, 2016; Volume 3.
51. Support Program Launched by the Polish Government, Consisting of a Subsidy of 1/3 of Solar Cell Costs for Households. 2019. Available online: https://www.gov.pl/web/energia/program-moj-prad-zalozenia-szczegolowe (accessed on 1 July 2020).
52. Amendments to the Excise Tax Act by the Provisions of the Act of 11 January 2018 on Electromobility and Alternative Fuels (Journal of Laws, Item 317, as Amended)—In Polish. Available online: http://prawo.sejm.gov.pl/isap.nsf/download.xsp/WDU20180000317/T/D20180317L.pdf (accessed on 1 July 2020).
53. Clean Air Program (In Polish: Czyste Powietrze), 2019. Available online: http://nfosigw.gov.pl/czyste-powietrze/o-programie-czyste-powietrze-/ (accessed on 16 July 2020).
54. Report of the Federation of Polish Food Banks—I Don't Waste Food (In Polish: Raport Federacji Polskich Banków Żywności—Nie Marnuję Jedzenia), 2018. Available online: https://bankizywnosci.pl/wp-content/uploads/2018/10/Przewodnik-do-Raportu_FPBZ_-Nie-marnuj-jedzenia-2018.pdf (accessed on 10 July 2020).
55. Patermann, C.; Aguilar, A. The origins of the bioeconomy in the European Union. *New Biotechnol.* **2018**, *40*, 20–24. [CrossRef]
56. Bioeconomy—Strategic Policy Direction of the European Union. (In Polish: Biogospodarka—Strategiczny kierunek polityki Unii Europejskiej, Kongres Ekoinwestycje w Przemyśle Spożywczym). 2017. Available online: http://bikotech.pl/wp-content/uploads/2017/02/Biogospodarka-Strategiczny-kierunek-polityki-Unii-Europejskiej-dr-hab.-in%C5%BC.-Monika-%C5%BBubrowska-Sudo%C5%82.pdf (accessed on 16 July 2020).
57. Adamowicz, M. Bioeconomy-Concept, Application and Perspectives. *Probl. Agric. Econ.* **2017**, *1*, 29–49.
58. Tsui, T.H.; Wong, J.W. A critical review: Emerging bioeconomy and waste-to-energy technologies for sustainable municipal solid waste management. *Waste Dispos. Sustain. Energy* **2019**, *1*, 151–167. [CrossRef]
59. Fuentes-Saguar, P.; Mainar-Causapé, A.; Ferrari, E. The role of bioeconomy sectors and natural resources in EU economies: A social accounting matrix-based analysis approach. *Sustainability* **2017**, *9*, 2383. [CrossRef]
60. Haarich, S. *Bioeconomy Development in EU Regions: Mapping of EU Member States'/Regions' Research and Innovation Plans & Strategies for Smart Specialisation (RIS3) on Bioeconomy*; Publications Office of the European Union: Luxembourg, 2017.
61. Avdiushchenko, A. Challenges and opportunities of circular economy implementation in the Lesser Poland Region (LPR). In Proceedings of the 13th International Conference on Waste Management and Technology (ICWMT 13): Overall Control of Environmental Risks, Naxos Island, Greece, 13–16 June 2018.
62. Bocken, N.; Strupeit, L.; Whalen, K.; Nußholz, J. A review and evaluation of circular business model innovation tools. *Sustainability* **2019**, *11*, 2210. [CrossRef]
63. Short, M. Taking back the trash: Comparing European extended producer responsibility and take-back liability to us environment policy and attitudes. *Vand. J. Transnatl. Law* **2004**, *37*, 1217.
64. Sachs, N. Planning the funeral at the birth: Extended producer responsibility in the European Union and the United States. *Harv. Envtl. L. Rev.* **2006**, *30*, 51.
65. Pak, P. Haste makes e-waste: A comparative analysis of how the United State should approach the growing e-waste threat. *Cardozo J. Int. Comp. Law* **2008**, *16*, 241.
66. McCrea, H. Germany's Take-Back Approach to Waste Management: Is There a Legal Basis for Adoption in the United States. *Geo. Int. Environ. Law Rev.* **2010**, *23*, 513.
67. Fehm, S. From iPod to e-Waste: Building a Successful Framework for Extended Producer Responsibility in the United States. *Public Contract Law J.* **2011**, *41*, 173.
68. Wu, H.H. Legal development in sustainable solid waste management law and policy in Taiwan: Lessons from comparative analysis between EU and US. *NTU Law Rev.* **2011**, *6*, 461.
69. Gu, F.; Guo, J.; Hall, P.; Gu, X. An integrated architecture for implementing extended producer responsibility in the context of Industry 4.0. *Int. J. Prod. Res.* **2019**, *57*, 1458–1477. [CrossRef]
70. Gaines, S.E. The polluter-pays principle: From economic equity to environmental ethos. *Tex. Int. Law J.* **1991**, *26*, 463.

71. European Parliament and Council Directive (UE) 2018/851, 2018 (In Polish: Dyrektywa Parlamentu Europejskiego i Rady (UE) 2018/851). Available online: https://eur-lex.europa.eu/legal-content/PL/TXT/PDF/?uri=CELEX:32018L0851&from=EN (accessed on 1 July 2020).
72. Resolution No. 88 of the Council of Ministers of 1 July 2016 regarding the National Waste Management Plan 2022 (In Polish: Uchwała nr 88 Rady Ministrów z dnia 1 lipca 2016 r. w sprawie Krajowego planu gospodarki odpadami 2022). Available online: http://prawo.sejm.gov.pl/isap.nsf/DocDetails.xsp?id=WMP20160000784 (accessed on 1 July 2020).
73. Lewandowska, A. Economics of waste management in poland in the context of regional ecologization as exemplified by kujawsko-pomorskie voivodship. *Electron. J. Pol. Agric. Univ.* **2018**, *21*. [CrossRef]
74. Bogdan, W.; Boniecki, D.; Labaye, E.; Marciniak, T.; Nowacki, M. *Poland 2025: Europe's New Growth Engine*; McKinsey&Company: New York, NY, USA, 2015. Available online: http://mckinsey.pl/wp-content/uploads/2015/10/Poland-2025_full_report.pdf (accessed on 11 March 2017).
75. Report of the Central Statistical Office on Environmental Protection (in Poland—GUS), Environment 2018. Technical Report, Statistics Poland, 2018. Available online: https://stat.gov.pl/en/topics/environment-energy/environment/environment-2018,1,10.html (accessed on 18 September 2020).
76. Polish Waste Act (In Poland: Ustawa z dnia 14 grudnia 2012 r. o odpadach, art. 3, ust.1, pkt.7, 2012. Available online: http://isap.sejm.gov.pl/isap.nsf/DocDetails.xsp?id=WDU20130000021 (accessed on 18 September 2020)
77. Eurostat Report—Municipal Waste Statistics 2017 (Published in 2019). Available online: https://ec.europa.eu/eurostat/statistics-explained/index.php/Municipal_waste_statistics (accessed on 28 June 2020).
78. Gołek-Schild, J. *Installations for Thermal Treatment of Municipal Waste in Poland—An Energy Source of Environmental Significance*; The Bulletin of The Mineral and Energy Economy Research Institute of the Polish Academy of Sciences: Krakow, Poland, 2018.
79. Kulkarni, B.N.; Anantharama, V. Repercussions of COVID-19 pandemic on municipal solid waste management: Challenges and opportunities. *Sci. Total Environ.* **2020**, 140693. [CrossRef]
80. Selva, N.; Chylarecki, P.; Jonsson, B.G.; Ibisch, P.L. Misguided forest action in EU Biodiversity Strategy. *Science* **2020**, *368*, 1438–1439.
81. Hahladakis, J.N.; Purnell, P.; Iacovidou, E.; Velis, C.A.; Atseyinku, M. Post-consumer plastic packaging waste in England: Assessing the yield of multiple collection-recycling schemes. *Waste Manag.* **2018**, *75*, 149–159. [CrossRef]
82. Fischer, C. *Municipal Waste Management in Poland*; Technical Report; European Environment Agency: Copenhagen, Denmark, 2013.
83. Towards a circular economy—Waste Management in the EU. Available online: https://www.europarl.europa.eu/RegData/etudes/STUD/2017/581913/EPRS_STU%282017%29581913_EN.pdf (accessed on 28 August 2020).
84. Marino, A.; Pariso, P. Comparing European countries' performances in the transition towards the Circular Economy. *Sci. Total Environ.* **2020**, *729*, 138142. [CrossRef] [PubMed]
85. Wielgosiński, G.; Namiecińska, O. Municipal waste incinerators—The perspective of 2020. *Nowa Energy* **2016**, *2*, 1–15.
86. Waste Management (In Polish: Gospodarka Odpadami. Available online: http://poznan.wios.gov.pl/monitoring-srodowiska/publikacje/raport2017/08_Gospodarka_odpadami.pdf (accessed on 25 June 2020).
87. State Aid: Commission Approves Euro 64 Million Support for a Waste-To-Energy Highly Efficient Cogeneration Plant in Poland. Available online: https://ec.europa.eu/info/news/state-aid-commission-approves-eu64-million-support-waste-energy-highly-efficient-cogeneration-plant-poland-2019-oct-21_en (accessed on 25 June 2020).
88. About 1.2 million Ukrainians "Pass Through" the Polish Labor Market (In Polish: Przez polski Rynek Pracy "Przewija Się" ok. 1,2 mln Ukraińców). Available online: https://businessinsider.com.pl/wiadomosci/ukraincy-ilu-pracuje-w-polsce-dane-nbp/r1yf4pe (accessed on 1 July 2020).
89. Integrated Pollution Prevention and Reduction (IPPC), Reference document for the best available techniques for waste incineration, August 2006 (In Polish: Zintegrowane Zapobieganie i Ograniczanie Zanieczyszczeń (IPPC), Dokument Referencyjny dla najlepszych dostępnych technik dla spalania odpadów, Sierpień 2006). Available online: https://ippc.mos.gov.pl/ippc/custom/spalanie%20odpad%C3%B3w.pdf (accessed on 10 July 2020).

90. Environmental Protection 2018 (In Polish: Ochrona środowiska 2018). Available online: https://stat.gov.pl/obszary-tematyczne/srodowisko-energia/srodowisko/ochrona-srodowiska-2018,1,19.html (accessed on 25 June 2020).
91. Information of the Minister of the Environment on Municipal Waste Incineration Plants and Their Place in the Waste Management System (In Polish: Informacja Ministra Środowiska na temat spalarni odpadów komunalnych i ich miejsca w systemie gospodarki odpadami). Available online: https://odpady.net.pl/wp-content/uploads/2019/04/Informacja-Ministra-%C5%9Arodowiska-na-temat-spalarni-odpad%C3%B3w-komunalnych.pdf (accessed on 10 July 2020).
92. Dennison, G.; Dodd, V.; Whelan, B. A socio-economic based survey of household waste characteristics in the city of Dublin, Ireland. I & II. Waste composition. *Resour. Conserv. Recycl.* **1996**, *17*, 227–257.
93. Daskalopoulos, E.; Badr, O.; Probert, S. Municipal solid waste: A prediction methodology for the generation rate and composition in the European Union countries and the United States of America. *Resour. Conserv. Recycl.* **1998**, *24*, 155–166. [CrossRef]
94. Bandara, N.J.; Hettiaratchi, J.P.A.; Wirasinghe, S.; Pilapiiya, S. Relation of waste generation and composition to socio-economic factors: A case study. *Environ. Monit. Assess.* **2007**, *135*, 31–39. [CrossRef] [PubMed]
95. Gellynck, X.; Jacobsen, R.; Verhelst, P. Identifying the key factors in increasing recycling and reducing residual household waste: A case study of the Flemish region of Belgium. *J. Environ. Manag.* **2011**, *92*, 2683–2690. [CrossRef]
96. The Richer the Society, the More Garbage It Produces. (In POlish: Im Społeczeństwo jest Bogatsze, tym Więcej śmieci Produkuje), 2019. Available online: https://serwisy.gazetaprawna.pl/samorzad/artykuly/1417088,recykling-smieci-rosnie-ilosc-odpadow.html (accessed on 2 July 2020).
97. Rozkrut, D. Migration Statistics in Poland—Data and Methodology. Available online: http://www.instat.gov.al/media/3875/konf_migracioni_statstics-poloni.pdf (accessed on 2 July 2020).
98. Zambrano-Monserrate, M.A.; Ruano, M.A.; Sanchez-Alcalde, L. Indirect effects of COVID-19 on the environment. *Sci. Total Environ.* **2020**, 138813. [CrossRef]
99. Problems with Waste Management—A Letter to the Minister (In Polish: Problemy Gospodarki Odpadami—List Do Ministra). 2019. Available online: https://sozosfera.pl/odpady/problemy-gospodarki-odpadami-list-do-ministra/ (accessed on 2 July 2020).
100. Cyranka, M.; Jurczyk, M.; Pająk, T. Municipal Waste-to-Energy plants in Poland–current projects. E3S Web of Conferences. *EDP Sci.* **2016**, *10*, 00070.
101. Smol, M.; Kulczycka, J.; Kowalski, Z. Sewage sludge ash (SSA) from large and small incineration plants as a potential source of phosphorus–Polish case study. *J. Environ. Manag.* **2016**, *184*, 617–628. [CrossRef]
102. Ashraf, M.S.; Ghouleh, Z.; Shao, Y. Production of eco-cement exclusively from municipal solid waste incineration residues. *Resour. Conserv. Recycl.* **2019**, *149*, 332–342. [CrossRef]
103. Mysior, M.; Bialowiec, A.; Stępień, P., Technical and technological problems as well as application potential of waste torification (In Polish: Problemy techniczne i technologiczne oraz potencjał aplikacyjny toryfikacji odpadów). In *Innovations in Waste Management (In Polish: Innowacje w Gospodarce Odpadami)*; Wrocław University of Environmental and Life Sciences Publishing House (In Polish: Wydawnictwo Uniwersytety Przyrodniczego we Wrocławiu): Wrocław, Poland, 2018; pp. 59–78.
104. Triyono, B.; Prawisudha, P.; Aziz, M.; Pasek, A.D.; Yoshikawa, K. Utilization of mixed organic-plastic municipal solid waste as renewable solid fuel employing wet torrefaction. *Waste Manag.* **2019**, *95*, 1–9. [CrossRef] [PubMed]
105. Shankar Tumuluru, J.; Sokhansanj, S.; Hess, J.R.; Wright, C.T.; Boardman, R.D. A review on biomass torrefaction process and product properties for energy applications. *Ind. Biotechnol.* **2011**, *7*, 384–401. [CrossRef]
106. Kuo, W.C.; Lasek, J.; Słowik, K.; Głód, K.; Jagustyn, B.; Li, Y.H.; Cygan, A. Low-temperature pre-treatment of municipal solid waste for efficient application in combustion systems. *Energy Convers. Manag.* **2019**, *196*, 525–535. [CrossRef]

MDPI
St. Alban-Anlage 66
4052 Basel
Switzerland
Tel. +41 61 683 77 34
Fax +41 61 302 89 18
www.mdpi.com

Energies Editorial Office
E-mail: energies@mdpi.com
www.mdpi.com/journal/energies

www.ingramcontent.com/pod-product-compliance
Lightning Source LLC
LaVergne TN
LVHW080502180726
843515LV00016B/894